Other Titles in This Series

Mathematical World • Volume 6

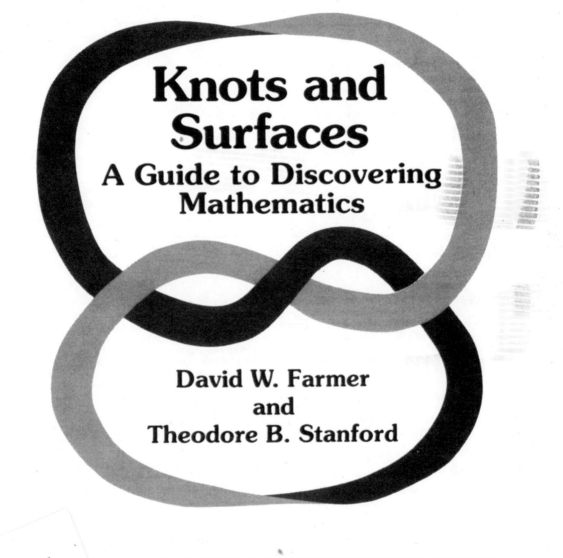

Knots and Surfaces

A Guide to Discovering Mathematics

David W. Farmer
and
Theodore B. Stanford

American Mathematical Society

2000 *Mathematics Subject Classification*. Primary 57–01; Secondary 05–01.

Library of Congress Cataloging-in-Publication Data

Farmer, David W., 1963–
    Knots and surfaces : a guide to discovering mathematics / David W. Farmer, Theodore B. Stanford.
    p. cm. — (Mathematical world, ISSN 1055-9426; v. 6)
    Includes bibliographical references.
    ISBN 0-8218-0451-0 (alk. paper)
    1. Graph theory. 2. Knot theory. 3. Surfaces. I. Stanford, Theodore B., 1964– . II. Title.
III. Series.

QA166.F37   1995
511′.5—dc20
                          95-30897
                              CIP

ISBN-13: 978-0-8218-0451-3

Table of Contents

Preface

This book is a guide to discovering mathematics.

Every mathematics textbook is filled with results and techniques which once were unknown. The results were discovered by mathematicians who experimented, conjectured, discussed their work with others, and then experimented some more. Many promising ideas turned out to be dead–ends, and lots of hard work resulted in little output. Often the first progress was the understanding of some special cases. Continued work led to greater understanding, and sometimes a complex picture began to be seen as simple and familiar. By the time the work reaches a textbook, it bears no resemblance to its early form, and the details of its birth and adolescence have been lost. The precise and methodical exposition of a typical textbook is often the first contact one has with the topic, and this leads many people to mistakenly think that mathematics is a dry, rigid, and unchanging subject.

We believe that the most exciting part of mathematics is the process of invention and discovery. The aim of this book is to introduce that process to you, the reader. By means of a wide variety of tasks, this book will lead you to discover some real mathematics. There are no formulas to memorize. There are no procedures to follow. By looking at examples, searching for patterns in those examples, and then searching for the reasons behind those patterns, you will develop your own mathematical ideas. The book is only a guide; its job is to start you in the right direction, and to bring you back if you stray too far. The discovery is left to you.

This book is suitable for a one semester course at the beginning undergraduate level. There are no prerequisites. Any college student interested in discovering the beauty of mathematics can enjoy a course taught from this book. An interested high school student will find this book to be a pleasant introduction to some modern areas of mathematics.

While preparing this book we were fortunate to have access to excellent notes taken by Hui–Chun Lee. We thank Klaus Peters and Gretchen Wright for helpful comments on an early version of this book.

David W. Farmer
Theodore B. Stanford
September, 1995

1

Networks

1.1 Countries of the insect world. Imagine a world populated by semi–intelligent insects. The world of the insects is divided into small countries, each country consisting of a few cities connected by dark narrow tunnels. In the course of their work and leisure the insects slowly walk these tunnels, and by the time they reach adulthood all insects know how their country's cities are connected. If an insect needs to travel from one city to another, and those cities are directly connected, then the connecting tunnel is taken. Maybe the route could be shortened by taking two short tunnels through another city, but the insects are only semi–intelligent, so this possibility never occurs to them. And the insects are poor at measuring distances, so they probably couldn't identify a shorter route even it they looked for it. Life on the insect world is calm and uneventful, the citizens blissfully bumping along in the dark, performing their chores with calm inefficiency.

Let's take a closer look at the world of the insects. Here are two insect countries:

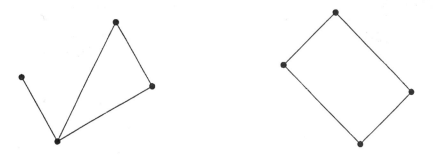

Our view from the 'outside' provides us with a complete picture of both countries. The insects are confined to the cities and tunnels, so they must expend more effort to get an accurate view of the layout. Suppose that communication between insect countries takes place by radio. Citizens from the above countries were talking, and they began to wonder if their two countries are the same. How can they determine that their countries have different layouts? First they observe that both countries have four cities and four main tunnels. So far, their countries appear similar. Then one says, "We have a city with just one tunnel leading to it." The other one says, "AHA! All our cities have two tunnels connected to them, so our countries are not set up the same way."

1

There are many other ways the insects could determine that their countries have different layouts. For example, each of these descriptions applies to exactly one of the countries above:

"My country has a city which connects directly to every other city."

"In my country, you can travel a route of four different tunnels and end up back where you started."

"In my country, you can travel a route of three different tunnels and end up back where you started."

Since the insects are bad at measuring distances, they are not always able to distinguish between layouts which we would see as different.

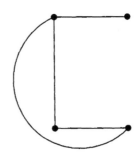

Task 1.1.1: Explain why the insects cannot distinguish between this country's layout and the first one shown previously.

Task 1.1.2: For each pair of countries, determine whether the insects would view them as the same or different. For those that are different, describe how the insects can tell them apart. Note: for each pair, the number of cities is the same and the number of tunnels is the same. If this were not the case, then the insects could immediately tell that the two countries had a different layout.

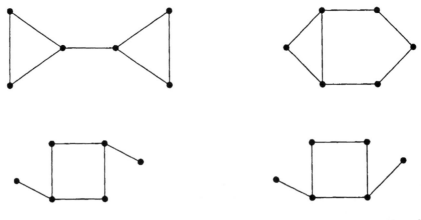

continued...

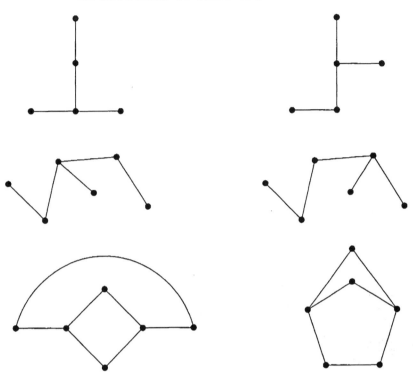

Task 1.1.3: Devise a precise description of what it means for two countries to be 'the same' as far as the insects are concerned.

Task 1.1.4: An insect says, "My country has seven cities and nine tunnels. One city has just one tunnel connected to it, one city has five tunnels connected to it, two cities have three tunnels connected to them, and the other three cities have two connecting tunnels." Draw two different countries which fit that description, and explain how the insects can tell them apart. How many different countries fit that description?

Advice. As you go through this book, you may find it helpful to keep a record of your thoughts and ideas. Set aside a notebook for this purpose. Put all of your work there, not just the final answers. It is important to keep a record of the entire process you went through as you worked on a problem, including work which didn't seem to lead to an answer. Your failed method on one problem could turn out to be the correct method for another problem. Having all your work in one place will help you see what you have done and will make it easy to find old work when you need it.

It is important that you spend sufficient time thinking about the Tasks as you encounter them. Some Tasks are easy and some are very difficult, so you should not expect to find a complete answer to every one. If a Task seems mysterious, it can help to discuss it with someone else. Occasionally you may skip a Task and come back to it later, but skipping a Task in the hope of finding the answers in the text will lead you nowhere. The only way for you to find an answer is to discover it yourself. Sometimes this will mean spending a long time on one Task. That is the nature of mathematical discovery. You will find that discovering your own mathematics is not at all like trying to learn mathematics which has already been discovered by someone else.

1.2 Notation, and a catalog

The ideas of the previous section fall under the mathematical topic of *graph theory*. The fanciful idea of insects crawling through dark tunnels will continue to be useful, but we will switch to using the mathematical terminology. Here is how to translate:

Insect name:	Math name:
country	graph
city	point or vertex
tunnel	line or edge

An example sentence is, "A graph is made up of points and lines." Note that 'vertices' is the plural of 'vertex,' so we can also say, "A graph consists of vertices connected by edges."

The actual picture we draw of a graph is called a **graph diagram**. Just as the insects could not distinguish between certain countries, the same graph can be represented by many different graph diagrams. The only important feature of the graph is how the various vertices are connected. Each graph diagram will have additional features, such as the lengths of the edges and the relative position of the vertices, but these aspects of the diagram have nothing to do with the graph itself.

Here are three diagrams of the same graph:

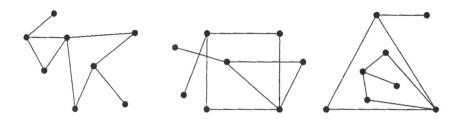

A diagram may appear to show two edges crossing, but if there is not a vertex at the junction then the edges do not actually meet. Think of it as two insect tunnels which pass each other but do not intersect. The topic of drawing graphs without crossing edges will be explored in a later section.

A graph is called **connected** if we can get from any vertex to any other vertex by traveling along edges of the graph. The opposite of connected is **disconnected**.

This can be thought of as a
disconnected graph with 9
vertices, or as two separate
connected graphs.

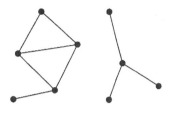

Any graph is just a collection of connected graphs; these are called the **components** of the graph.

The graphs we have been studying are presented as drawings on paper. It is easy to invent graphs which are described in other ways. For example, we can make a graph whose vertices are all of the tennis players in the world, and where an edge connects two players if they have played tennis together. We have defined a graph, although it would not be feasible to actually draw it. Another graph can be made by letting the vertices be the countries of the world, and having an edge connect two countries if those countries border each other. With the help of a map it would be possible to draw this graph. It is amusing to invent fanciful graphs and then try to determine what properties they have. For example, is the tennis player graph connected? If it is, that would mean each tennis player has played someone who has played someone who has . . . played Jimmy Connors. The play *Six Degrees of Separation* mentions, informally, the graph whose vertices are all of the people in the world, with edges connecting people who know each other. The title of the play comes from speculation that you can get from any one vertex to any other vertex by crossing at most 6 edges.

In order to make an organized study of graphs, we must impart a few more rules. Usually we do not allow our graphs to have more than one edge connecting two vertices, and we do not allow an edge to connect a vertex to itself.

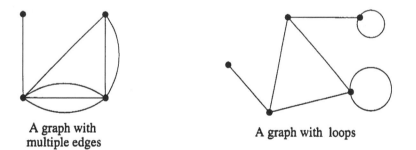

A graph with
multiple edges

A graph with loops

Unless we state otherwise, a 'graph' is a 'graph without loops or multiple edges.'

We classify graphs according to how many vertices they have. Here is a catalog of all graphs with 4 vertices:

You should convince yourself that the list is complete.

Task 1.2.1: Make a catalog of all graphs with 5 vertices.   Hint: there are between 30 and 40 of them. First find all the ones with no edges, then 1 edge, then 2 edges, and so on.

In the above Task it is difficult to be absolutely sure that you found all the graphs. Fortunately, there is something we can do to increase our confidence. For the graphs with 4 vertices we found a total of $1 + 1 + 2 + 3 + 2 + 1 + 1 = 11$ graphs, where we counted the graphs according to how many edges they have. Notice that the numbers form a symmetric pattern.

Task 1.2.2: Do your numbers from Task 1.2.1 form a symmetric pattern? If not, go back and fix your list. After your list is correct, explain why the symmetric pattern appears.

Task 1.2.3: Devise a code for describing a graph over the telephone. Note: your code only needs to describe a *graph*, not a *graph diagram*.

1.3 Trees

If we think of a graph as a roadmap then it is natural to look at the various routes we can take through the graph. A **path** in a graph is a sequence of edges, where successive edges share a vertex. To make things easier to read, we will describe a path by showing which vertices the path visits; this should not cause any confusion.

Example paths:

    (d, b, c, d)

    (f, e, d, b, c)

    (e, g, b, a, b, g)

A path is **closed** if it ends at the same vertex as it began. The first path above is closed. A path is **simple** if it doesn't use the same edge more than once. The first two paths above are simple. A simple closed path is sometimes called a **circuit**. The first path above is a circuit, and so is *(b, c, d, e, g, b)*. A graph is **connected** if there is a path from any one vertex to any other vertex.

A graph is a **tree** if it is connected and it doesn't have any circuits. Here are three trees:

Task 1.3.1: What is the relationship between the number of vertices and the number of edges in a tree? Why does this relationship hold?

Trees are particularly simple kinds of graphs, so our plan is to study trees, and then to use trees to study other graphs. Here are all trees with 5 vertices:

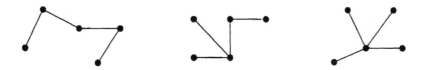

Those trees should be in your catalog of graphs from Task 1.2.1. Here are all trees with 6 vertices:

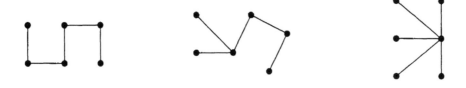

continued...

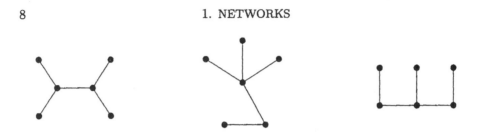

Task 1.3.2: Make a list of all trees with 7 vertices. If you feel ambitious, make a list of all trees with 8 vertices. Hint: there are between 20 and 30 of them.

Task 1.3.3: Suppose you had plenty of time and you wanted to make a list of all trees with a given large number of vertices; say, all trees with 12 vertices. Describe the method you would use. Is your method guaranteed to give a complete list with no repeats? Is your method practical?

Task 1.3.4: In Task 1.2.3 you devised a code for describing a graph over the telephone. Suppose that you only needed the code for describing trees. Is it possible to devise a simpler code which still works in this case?

1.4 Trees in graphs

A tree inside a graph which hits every vertex of the graph is called a **spanning tree**. A spanning tree must use the edges in the graph, and it must hit every vertex. A useful way to show a spanning tree is to highlight the edges in the tree:

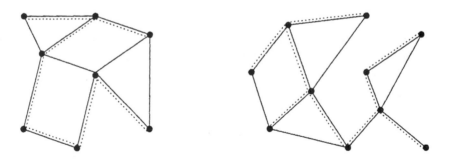

A graph can have many different spanning trees. Here are three different spanning trees for the same graph:

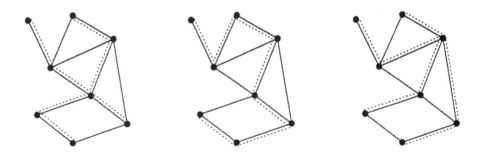

It is important to keep in mind that a graph can have several different spanning trees, so without a picture the term 'spanning tree' can be ambiguous.

Task 1.4.1: Devise a way of counting the number of spanning trees of a graph.

In the next section we use spanning trees to study graphs.

1.5 Euler's formula

A graph diagram divides the plane into separate regions:

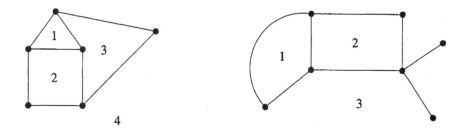

The first diagram divides the plane into 4 regions, and the second divides the plane into 3 regions. Note that the big outside area counts as a region.

Task 1.5.1: Draw several graphs and record the following information:
 – the number of vertices in the graph (call it v)
 – the number of edges in the graph (call it e)
 – the number of separate regions of the graph (call it f)
 – the number of vertices in a spanning tree (call it A)
 – the number of edges in a spanning tree (call it B)
 – the number of edges not in a spanning tree (call it C)

Here is an example. Find a spanning tree and check that the numbers are correct:

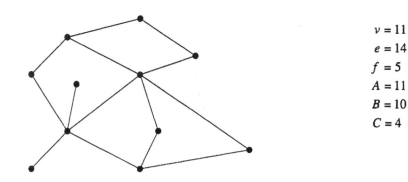

$v = 11$
$e = 14$
$f = 5$
$A = 11$
$B = 10$
$C = 4$

Note: for this Task you should only use connected graphs which you have drawn without crossing edges. A diagram drawn without crossing edges is called a

planar diagram. The importance of using planar diagrams in this Task is discussed in the next section.

Task 1.5.2: Look at the information you recorded and try to find patterns and relationships among the six quantities.

Task 1.5.3: Explain why the observations you made are correct. Note: one of your observations may have been $A = B + 1$. You already discussed this in Task 1.3.1.

Task 1.5.4: Explain why your observations can be used to show $v - e + f = 2$.

The equation $v - e + f = 2$ is known as **Euler's Formula**. It was first discovered by the Swiss mathematician Leonhard Euler in 1736. Note: Euler is pronounced 'Oiler.' Say it out loud a few times. This will keep you from looking foolish later.

Task 1.5.5: Suppose a graph has 7 vertices and 9 edges. Use Euler's formula to predict how many separate regions it would have if you drew the graph. Draw such a graph and check if your prediction is correct.

1.6 Planar graphs

Euler's formula $v - e + f = 2$ is true for any connected graph which is drawn without crossing edges. For instance:

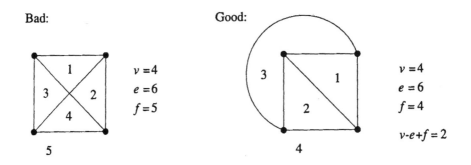

We say that a graph is **planar** if it has a diagram without crossing edges. Above are two diagrams of the same graph, but Euler's formula only works in the second case. This is usually expressed as "Euler's formula holds for connected planar graph diagrams."

Task 1.6.1: What can you say about $v - e + f$ if the graph is not connected?

The graph shown above is called 'the complete graph on 4 vertices,' and it is denoted K_4, pronounced "kay four." This means that it has 4 vertices and each vertex is connected to every other vertex. Similarly, K_5 is the graph with 5 vertices and each vertex is connected to every other vertex.

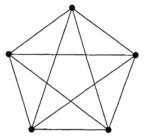

Here is a representation of K_5:

Task 1.6.2: How many edges does K_5 have? K_6? K_7?

Task 1.6.3: Explain why K_n has $1 + 2 + 3 + \cdots + (n - 1)$ edges. We will find another expression for this in Task 1.7.9.

Task 1.6.4: Try to draw K_5 without any crossing edges. Make at least four attempts.

After four attempts at Task 1.6.4, you should stop. Further attempts would be pointless because it is *impossible* to draw K_5 without any crossing edges. In other words, K_5 is not planar. We will use Euler's formula to show why the Task is impossible.

The reason we write f for the number of separate regions of a graph is that those regions are usually called **faces**. A key fact we need is that each edge of a planar graph diagram is a border of two faces.

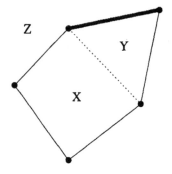

The dotted edge is a border of face X and face Y, and the fuzzy edge is a border of faces Y and Z.

The number of edges of a face is called the **order** of the face. In the diagram above, face X has order 4, face Y has order 3, and face Z has order 5.

An important relationship between the number of edges and the number of faces in a planar graph is:

$$3f \leq 2e.$$

We will use the concept of order, along with the observation that each edge borders two faces, to establish this inequality. Then we will use the inequality to show that K_5 does not have a planar diagram. But first, do this Task:

Task 1.6.5: Draw a few planar graph diagrams and check that $3f \leq 2e$ holds in each case. What graphs have $3f = 2e$?

The first step in showing $3f \leq 2e$ is this observation:

Counting Observation. If you add up the number of edges bordering every face, then you get twice the number of edges. In other words:

the sum of all the orders of the faces $= 2e$.

For example, in the diagram above there are three faces: X, Y, and Z. Adding up the orders of each faces gives $4 + 3 + 5 = 12$. And sure enough, 12 is twice the number of edges in the graph.

The reason behind the Counting Observation is that each edge is a border of two faces, so as we add up the number of edges around each face, each edge gets counted twice.

Now, we need at least 3 edges to make a face, so each face must have order at least 3. So, adding up the orders of each face gives something at least $3f$, so

$3f \leq$ the sum of all the orders of the faces,

so,

$$3f \leq 2e.$$

This is the inequality we wanted. We now use it to show K_5 is not planar.

The graph K_5 has 5 vertices and 10 edges. **IF** K_5 had a planar diagram, then Euler's formula $v - e + f = 2$ would tell us $f = 7$. The inequality we proved would then say

$$3f \leq 2e$$
$$3 \cdot 7 \leq 2 \cdot 10$$
$$21 \leq 20.$$

Of course, 21 is more than 20, so something is wrong. The error was the assumption that K_5 had a planar diagram. The inescapable conclusion is that K_5 does not have a planar diagram. In other words, it is impossible to draw K_5 with no crossing edges. In other other words, K_5 is not planar.

Task 1.6.6: For each case below, either draw a planar graph with the given information, or explain why this is not possible.

a) A graph with $v = 7$ and $e = 17$.

b) A graph with $v = 8$ and $e = 12$.

c) A graph with $v = 7$ and $e = 15$.

The method we used to show that K_5 does not have a planar diagram can also be used to show that some other graphs are not planar, but the method is not foolproof.

This graph is called $K_{3,3}$. It has 6 vertices and 9 edges, and it does not have a planar diagram.

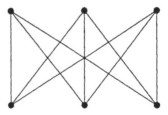

The name $K_{3,3}$, pronounced "kay three three," means the graph consists of one set of 3 vertices, each of which is connected to another set of 3 vertices. Here are more examples to explain the notation:

Task 1.6.7: How many edges does $K_{2,5}$ have? $K_{6,6}$? $K_{100,200}$? $K_{n,m}$?

Task 1.6.8: Try to draw $K_{3,3}$ without crossing edges. You will not succeed, because $K_{3,3}$ is not planar, but you should make a few attempts anyway.

We showed that K_5 does not have a planar diagram, but the same method doesn't work for $K_{3,3}$. Fortunately, we can modify the method. The key to the modification is this observation:

No Triangles Observation. $K_{3,3}$ does not contain any simple closed paths of three edges.

The quick way to say it is: $K_{3,3}$ does not contain any triangles. This means that if we could draw $K_{3,3}$ without any crossing edges, then every face would have order at least four.

Task 1.6.9: Explain why a planar graph with no triangles must have $4f \leq 2e$. Use this to show that $K_{3,3}$ does not have a planar diagram.

It turns out that understanding K_5 and $K_{3,3}$ is fundamental to understanding all nonplanar graphs. See the Notes at the end of the chapter for an explanation.

Task 1.6.10: Do all trees have a planar diagram?

1.7 Paths in graphs

In this section we study various kinds of paths in a graph. We start with a puzzle and two problems.

Task 1.7.1: Trace this figure without picking up your pencil and without repeating a line, or explain why this is impossible.

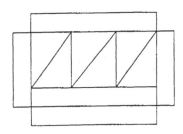

Task 1.7.2: A huge snowstorm has covered the town with snow! There is only one snowplow, and the roads need to be plowed as soon as possible. The snowplow driver figures that if she can manage to plow all the streets in one trip, without driving over a street which has previously been plowed, then this will get the job done quickly. Trace an efficient route on the map shown to the right, or explain why a completely efficient route is impossible.

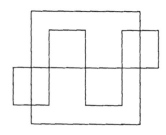

Task 1.7.3: The highway inspector must evaluate the safety of all the roads in town. Since he is lazy, he wants to travel along each road exactly once; he does not want to drive on a road which he has already inspected. Trace an efficient route for the lazy inspector, or explain why no such route is possible.

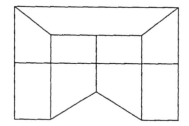

Hopefully you traced the figure in Task 1.7.1 and you also found an efficient path for the snowplow in Task 1.7.2. If not, then go back and try again! There is no efficient path for the lazy inspector in Task 1.7.3. No matter how hard you try, it is impossible to trace the roads of that town without using some road more than once. Our next goal is to give an explanation for this.

The three Tasks above have the same theme: each gives a graph and asks if there is a simple path which uses every edge in the graph. Recall that *simple* means that no edge is used twice. In a graph, a simple path which uses every edge is called an **Euler path**. An **Euler circuit** is an Euler path which begins and ends at the same vertex.

If you check back at your answers to the above Tasks, your solution to the puzzle is an Euler *path*, and your solution to the snowplow problem is an Euler *circuit*. The graph for the highway inspector has neither an Euler path nor an Euler circuit.

Task 1.7.4: Find an Euler path in each of these graphs. The Task will become very easy once you determine the significance of the * vertices.

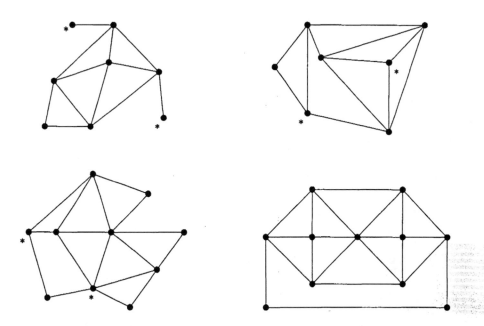

The key to Euler paths in a graph lies in careful examination of the vertices. The number of edges connected to a vertex is called the **order** of the vertex.

Each vertex in this graph is labeled with its order.

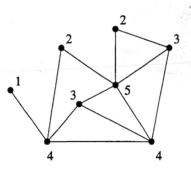

Task 1.7.5: What relationship exists between an Euler path in a graph and the orders of the vertices in that graph? If you don't see a relationship yet, then do more examples. Explain why your observations are correct.

The previous Task is the key point of this section. Be sure to give it sufficient thought.

Task 1.7.6: Repeat your work in the previous Task for Euler *circuits*.

Task 1.7.7: Draw a graph with exactly *one* odd–order vertex. Does it have an Euler path?

Task 1.7.7 is a trick question: it is impossible to draw a graph with exactly one odd–order vertex. Here is one way to see this:

Task 1.7.8: Explain why the Counting Observation following Task 1.6.5 is valid with 'face' replaced by 'vertex.' Use this New Counting Observation to explain why Task 1.7.7 is impossible.

Task 1.7.9: Use the New Counting Observation in Task 1.7.8 to show that K_n has $n(n-1)/2$ edges. Compare this to Task 1.6.3.

The *Handshake Principle* is a slight generalization of our rule that a graph cannot have exactly one odd–order vertex.

The Handshake Principle. Take a group of people and have each person shake hands with various other people in the group. The number of people who shook hands an odd number of times must be even.

In terms of a graph, the Handshake Principle says that the number of odd–order vertices must be even.

We have been studying paths which cross every edge once. Another interesting problem is to study paths which visit every vertex once. This idea first appeared in a game invented by the English mathematician Sir William Rowan Hamilton. He sold the idea to a game producer, but the game never made any money!

Hamilton's problem: find a path in this graph which visits every vertex once, and which ends at the same vertex as it began. In the original version each vertex was labeled with a famous city, and the object was to 'Travel the World.'

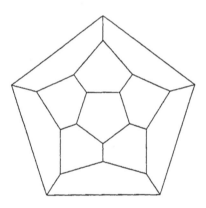

A **Hamiltonian path** in a graph visits every vertex exactly once. The path is a **Hamiltonian circuit** if it ends at the same vertex as it began. A Hamiltonian path can also be thought of as a spanning tree with no 'branches.' See the middle graph at the very end of Section 1.4 for an example.

The study of Hamiltonian paths is much more difficult than the study of Euler paths. A few rules are known, and any specific graph can be analyzed by computer, but nobody has found a simple method for determining when a graph has a Hamiltonian path.

1.8 Dual graphs

We introduce a way to take one graph diagram and use it to produce another graph. This new graph is called the *dual* of the original diagram.

Start with any planar diagram:

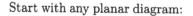

Put a vertex in each separate
region of the original diagram.

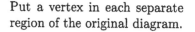

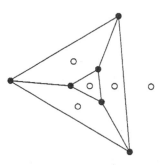

Connect the new vertices which
are in adjacent regions of the
original diagram. Each original
edge will have a new edge crossing it.

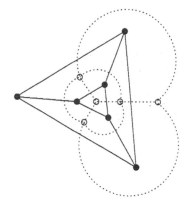

The graph made with the new
edges and vertices is called the
dual of the original graph diagram.

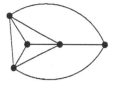

The above procedure is called **taking the dual** of the original graph diagram.
We will see that there are interesting relationships between a graph and its dual,
and the dual graph is useful for solving certain problems.

Recall that a *graph* is not permitted to have loops or multiple edges. Because of this restriction, the dual of a graph diagram might not be a 'graph' in the strictest sense. For example:

That dual graph has multiple edges, so it is not 'really' a graph. It is too much effort to keep worrying about this distinction, so for the rest of this section we will permit our graphs to have loops and multiple edges.

Task 1.8.1: Draw a few planar graph diagrams, and then find their duals.

Task 1.8.2: What is the relationship between v, e, f for a graph and v, e, f for its dual?

After doing Task 1.8.2, you should look back at the first part of Task 1.7.8.

Task 1.8.3: What is the relationship between a graph, its dual, and the dual of the dual?

Task 1.8.4: Find a few graphs which are the same as their dual.

Task 1.8.5: Can you draw a curve which crosses each edge of this figure exactly once? The curve is not allowed to cross itself. Two failed attempts are shown below.

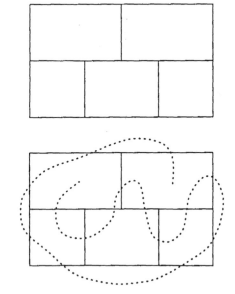

Task 1.8.6: Would the above Task be any easier if the curve was allowed to cross itself?

1.9 A map of the United States

On the next page is a map of the 48 continental US states. It might not look like a typical US map, but there is a way to label each region with the name of a state so that each state borders the same states as on the usual map.

Task 1.9.1: Label the 48 regions of the map. Explain how you determined which region corresponds to which state. Is there only one way to correctly label the regions?

Task 1.9.2: Which state borders the most other states?

Task 1.9.3: Devise a similar map for your favorite geographic region.

1.10 Coloring graphs

A cartographer is producing a new map of the world. To make the map easier to read, each country will be given a color, and adjacent countries must be assigned different colors. Two countries can have the same color provided they don't border each other. Here is a map which we have colored using the numbers 1, 2, 3 and 4:

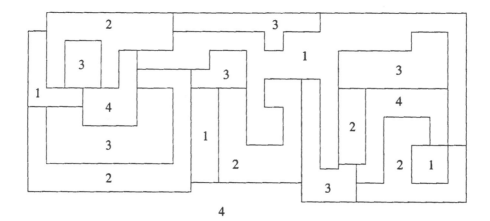

Note that the large outside region also gets a color.

There are a few rules about the 'maps' we will consider. Each region represents a separate country, and countries are not permitted to have 'satellite states.' For example, on the usual world map we would not permit Alaska to be part of the Unites States, because Alaska is not directly connected to the other states. We can use the world map as an example for our 'map,' but must permit Alaska to be a different color than the 48 continental states. Also, we do not consider two regions to be bordering if they have only one point in common. For example, Colorado and Arizona do not border each other on the United States map. Another example can be seen in the lower right portion of the example above. The small square labeled '1' does not border the large region labeled '1.'

Task 1.10.1: Draw a few maps and color them. Use as few colors as possible.

All of your maps from Task 1.10.1 can be colored with at most 4 colors. If you used 5 or more colors in some of your maps then go back and color them again. All maps can be colored with at most 4 colors. If you are skeptical about this then try to design a map that requires 5 colors.

The statement that any map can be 4–colored is known as "The Four–Color Theorem." The Four–Color Theorem is easy to understand, and trying a few examples makes it easy to believe, but giving a complete proof is extremely difficult. The Four–Color Theorem has been widely believed for more than 100 years, but a formal proof was not given until 1974. The proof was controversial because a computer was used for part of the calculation. Some people are still skeptical that all of the details were checked properly, and an independent calculation has not yet been completed. In this book we give a detailed proof that any graph can be 6–colored, and we roughly indicate why any graph can be 5–colored. If you figure out a simple proof of The Four–Color Theorem, then your name will be immortalized forever in the annals of mathematics.

We will spend the rest of this section using graphs to study map colorings. It is easy to associate a graph to a map: put a vertex in each region, including the large outside region, and connect adjacent regions with an edge. This works just like finding a dual graph. Here is the graph associated to the map shown previously:

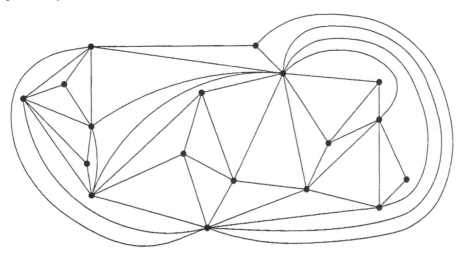

And here is a simpler example. Each vertex in the graph is colored the same as the corresponding region of the map.

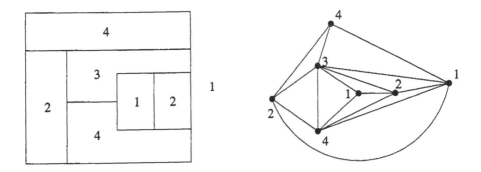

Coloring a map is now replaced by coloring a planar graph: we color the *vertices* of the graph so that adjacent vertices, vertices which share an edge, have different colors. For the purposes of coloring, the graph has the same information as the map. We will study graph coloring because then we can make use of our knowledge of planar graphs.

Task 1.10.2: An obvious way to try to get a map which cannot be 4–colored is to draw 5 countries, each one of which borders the other 4 countries. Since each country borders every other country, 5 different colors would be needed. However, such an arrangement is impossible. Explain why.

It might seem that since we can't have 5 countries simultaneously bordering each other, that would automatically imply that any map can be 4–colored. However, the logic is flawed. To see the flaw, look at the map below.

Each region borders only two other regions, but the map requires three colors.

The above example shows that the number of neighbors of each region is not always directly related to the number of colors needed to color the map. Another version of the same idea is given in the next Task.

Task 1.10.3: Color the vertices of each of the following graphs with as few colors as possible. Note that the number of colors is not directly related to the number of neighbors of each vertex.

For certain kinds of graphs we can say exactly how many colors are needed to color them.

Task 1.10.4: Suppose a graph has every vertex of order 3 or less. Explain how to 4–color that graph.

Task 1.10.5: Explain why any tree can be 2–colored.

Task 1.10.6: Suppose each separate region of a graph has an even number of sides. Is it necessarily true that the graph can be 2–colored? Note: you are coloring the *graph*, so each vertex gets a color, and vertices sharing an edge must get different colors.

Task 1.10.7: Construct a map by drawing a continuous curve which begins and ends at the same point and crosses itself as many times as you want. An example is shown below. Determine how many colors are needed to color such a map.

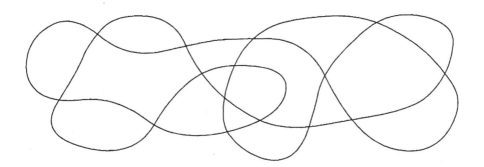

Task 1.10.8: Repeat Task 1.10.7 for maps made by several different overlapping curves. That is, draw several different curves as in Task 1.10.7 and permit the curves to cross each other.

Task 1.10.9: How many colors are needed to color K_n? Note: for $n \geq 5$ that graph will not be planar, but it still makes sense to color the vertices so that vertices sharing an edge get different colors.

Task 1.10.10: How many colors are needed to color $K_{n,m}$?

Task 1.10.11: Devise some graphs which require exactly three colors.

1.11 The six–color theorem

Since The Four–Color Theorem is so hard, we will prove The Six–Color Theorem: every graph can be colored with at most six colors. This is a common occurrence in mathematics: the ultimate goal may be out of reach, but we can still get satisfaction by proving a partial result. To prove The Six–Color Theorem, we need this fact about planar graphs.

Planar Graph Fact. Every planar graph has at least one vertex of order five or less.

In other words, we can't have every vertex of order six or more. We will prove The Six–Color Theorem, and after that we will prove the Planar Graph Fact. But first...

Task 1.11.1: Draw a graph where each vertex has order at least five. What does this say about the Planar Graph Fact?

We will describe a procedure for 6–coloring a planar graph, and then we will illustrate the procedure with an example.

To 6–color any planar graph, just follow these four steps:

Step 1: Locate a vertex of order 5 or less.

Step 2: Delete that vertex and all edges connected to it.

Keep repeating Steps 1 and 2 until only 5 vertices are left. Keep track of the order in which you deleted the vertices.

Step 3: Color the five remaining vertices with the colors 1, 2, 3, 4, 5.

Step 4: Put back the last vertex and edges you deleted. Color that vertex a different color than the vertices adjacent to it.

Repeat Step 4, replacing vertices in the reverse order they were deleted. After you put all the vertices back you will have reconstructed the original graph and each vertex will be colored with one of the 6 colors.

The Planar Graph Fact mentioned above is what makes the procedure work. Since a planar graph must have a vertex of order 5 or less, Step 1 can always be done. As you put the vertices back, each vertex is connected to at most 5 other vertices. Since there are 6 colors available, Step 4 can always be done.

Here is an example of using the procedure to 6–color a graph. The important thing to notice is that the procedure described above is followed exactly. No cleverness is needed. We just mechanically follow the plan and everything will work out right. First remove vertices one–by–one, also removing the connecting edges. If there is more than one vertex of order less than 6 then it doesn't matter which one we choose. Stop when there are only 5 vertices left.

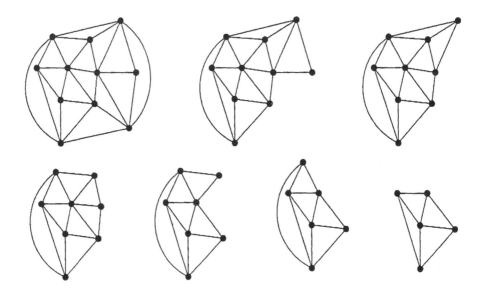

Now color the remaining vertices, and then replace the vertices in the reverse order they were deleted. When we replace a vertex we color it with any number different from the vertices it is adjacent to.

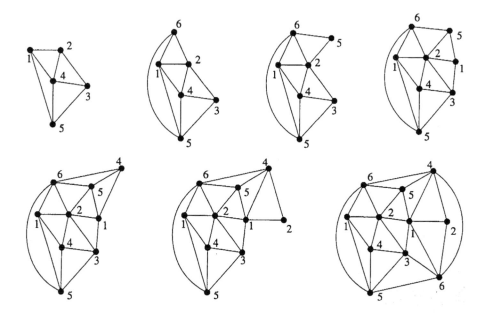

The result is a 6–coloring of the graph.

Of course, it is also possible to 4–color that same graph, but this requires cleverness. Nobody has ever found a simple procedure which is guaranteed to 4–color any planar graph.

Task 1.11.2: Four–color the graph shown above.

Now we prove the planar graph fact used above. This is the plan: we assume that it is possible for a planar graph to have all vertices of order at least 6. Using this assumption we will end up with a nonsensical statement. This shows that the assumption was not valid.

We use Euler's formula $v - e + f = 2$. We also need the earlier observations:

Sum of the orders of all the faces $= 2e$,

and

Sum of the orders of all the vertices $= 2e$.

First consider the faces. Each face must have at least 3 edges. In other words, the order of each face must be ≥ 3. If we counted '3' for each face then we would get something smaller than if we counted the order of each face. In other words, $3f \leq 2e$. This can also be written $f \leq \frac{2}{3}e$. Note: a more complete version of this same argument is given in Section 1.6.

Next consider the vertices. **IF** every vertex has order at least 6, then counting '6' for each vertex gives something smaller than counting the order of each vertex. In other words, $6v \leq 2e$. This can also be written $v \leq \frac{1}{3}e$.

Here are the three relationships we have:

$$2 = v - e + f$$

$$v \le \frac{1}{3}e$$

$$f \le \frac{2}{3}e.$$

Putting them all together gives

$$2 \le \frac{1}{3}e - e + \frac{2}{3}e$$

so,

$$2 \le 0.$$

That last inequality is nonsense, so our assumption that all vertices had order at least 6 is invalid, so we conclude that at least one vertex has order 5 or less. End of Proof.

Task 1.11.3: Modify the above calculation to show that if a planar graph has all vertices of order 5 or more, then the graph must have at least 30 edges. Conclude that it must have at least 12 vertices. Find a graph with 12 vertices, each vertex having order 5.

We end this chapter by describing a procedure for 5–coloring any planar graph. This procedure is a modification of the 6–coloring method. The basis for the improvement is the following fact:

Another Planar Graph Fact. Given any 5 vertices in a planar graph, there must be two which are not directly connected to each other.

The proof is simple: If all 5 vertices were directly connected to each other then that would be K_5, but K_5 is not planar.

We use this new fact to modify Step 2 of the procedure. We delete vertices and edges as before, but then we *glue two of the vertices together*. Specifically, if there were five vertices connected to the vertex we just deleted, choose two of them which are not adjacent, and then form a new graph by gluing those two vertices together. If this gives multiple edges then delete all but one of each repeated edge. This ends the new Step 2. Here is an example using the previous graph:

Delete a vertex of
order 5 or less.

The vertices marked * are not adjacent, so we can glue them together.

Combine multiple edges.

In the new procedure, repeat the new Step 2 until there are only 4 vertices left. For the new Step 3, color each of the remaining 4 vertices a different color. For the new Step 4, replace the deleted vertices in the reverse order they were removed. This is slightly more complicated than the original method because sometimes you have to rip apart two vertices which had been glued together.

To complete this Task, convince yourself that the procedure works. Note that if the graph has a vertex of order 4 or less then we can choose to delete that vertex first, so we only need to use the "gluing two vertices" step when every vertex has order 5 or more.

1.12 Notes

Note 1.12.a: Graphs were invented by Leonhard Euler to solve the 'Seven bridges of Königsberg' problem. The city of Königsberg, now known as Kaliningrad, Russia, had seven bridges. Here is what it looked like in Euler's time:

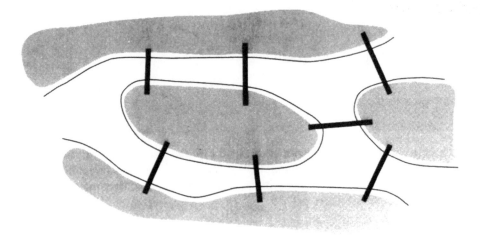

The citizens of the city wondered about the following question: Was it possible to take a walk which crossed each of the seven bridges exactly once?

It became a common recreational activity to try and complete such a walk, but nobody succeeded in the task. Euler showed that the task was impossible. His method used the idea of an Euler path in a graph.

A walk crossing each bridge exactly once would correspond to an Euler path in this graph:

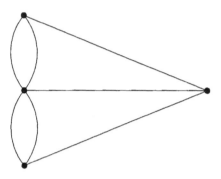

Each vertex corresponds to a piece of land, each edge corresponds to a bridge, and an Euler path in the graph corresponds to a walk crossing each bridge exactly once. Since the graph has four odd–order vertices, there is no Euler path, so it is impossible to walk across each of the seven bridges exactly once.

Note 1.12.b: There is a large class of graphs which we have not discussed. These are known as 'directed graphs,' commonly called **digraphs**. In a digraph the edges are like one-way streets. Here are two examples:

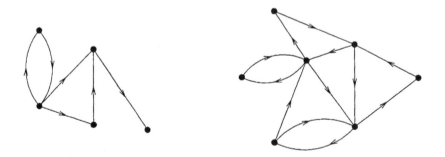

The main difference between a graph and a digraph is that in a digraph a path must 'follow the arrows.' For example, if you ignore the arrows then both graphs above have Euler paths. But if you must follow the arrows, then only the digraph on the right has an Euler path.

Note 1.12.c: Some people use the word **valence** where we use the word *order*.

Note 1.12.d: A graph contained within another graph is called a **subgraph**. This means that the vertices and edges of the subgraph are also vertices and edges of the larger graph. A spanning tree is a special type of subgraph.

Many properties of a graph also hold for all its subgraphs. For example, a subgraph of a planar graph is also planar. Also, if a graph can be colored with at most N colors, then so can all its subgraphs.

Note 1.12.e: The usual way to prove Euler's formula $v - e + f = 2$ is by induction on the number of vertices and edges in the graph.

Note 1.12.f: We showed that K_5 and $K_{3,3}$ are not planar. An important result known as "Kuratowski's Theorem" says that all nonplanar graphs 'contain' either K_5 or $K_{3,3}$, or both. Exactly what it says is:

Kuratowski's theorem. Starting with any nonplanar graph, you can produce one of K_5 or $K_{3,3}$ by repeatedly performing these three moves:

Move A: Delete an edge.

Move B: Delete a vertex and all edges connected to it.

Move C: Delete a vertex of order 2, combining the two 'dangling' edges into one edge.

The idea of Kuratowski's theorem is that we can take a nonplanar graph and throw away a bunch of it until we reduce down to either K_5 or $K_{3,3}$. The proof can be found in [GT], or in any other good introductory graph theory book.

Kuratowski's theorem says that if a graph is nonplanar then it 'contains' either K_5 or $K_{3,3}$. The reverse is also true: if a graph 'contains' either K_5 or $K_{3,3}$ then it is nonplanar. Is this obvious?

Task 1.12.1: Determine if the following statement is true:

Suppose a graph is nonplanar, but deleting any one edge results in a graph that *is* planar. Then the original graph must be either K_5 or $K_{3,3}$.

Either explain why the statement is true, or modify the statement so that it becomes true.

Task 1.12.2: The Petersen graph is nonplanar. Draw a picture to show that the Petersen graph contains one of K_5 or $K_{3,3}$.

The Petersen graph.

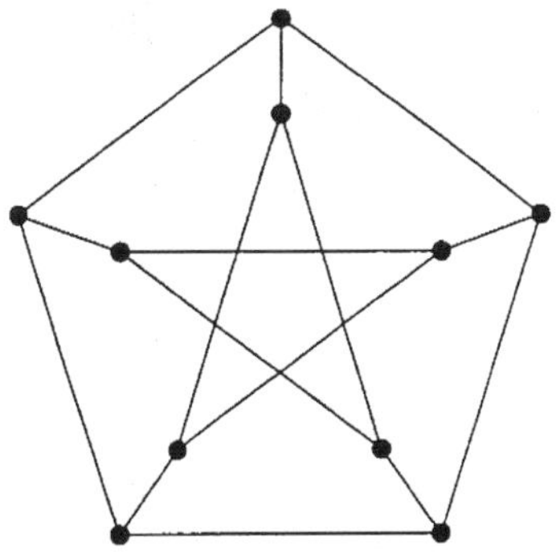

Task 1.12.3: Here is an amusing way to show that the Petersen graph is nonplanar. First, show that the Petersen graph does not contain any simple closed paths with fewer than 5 edges. Second, explain why this means that any dia-

gram of the Petersen graph would have $5f \leq 2e$. Finally, use $v - e + f = 2$ and $5f \leq 2e$ to show that the Petersen graph is nonplanar.

Note 1.12.g: In this chapter we discovered that if a graph has an Euler path then it must have 0 or 2 odd order vertices. The question remains: if a graph has 0 or 2 odd order vertices, does that automatically imply that it has an Euler path? The answer is 'yes', and this can be proved by induction on the number of vertices in the graph. See [GT] or any other introductory book on graph theory.

Task 1.12.4: Which graphs K_n and $K_{n,m}$ have an Euler path? Euler circuit?

Task 1.12.5: Use your answer to Task 1.12.4 to show that it is possible to place the tiles from the game of Dominoes in a circle so that the number of spots on the end of each tile matches the number of spots on the end of the adjacent tile.

Note 1.12.h: When discussing planar graphs we made the point of saying 'dual of a planar *diagram*.' This is important because we use the diagram to find the dual, and a planar graph can have several different planar diagrams. Here are two different diagrams of the same graph:

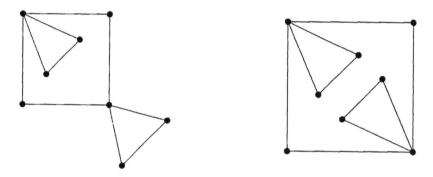

And here are the duals of those diagrams:

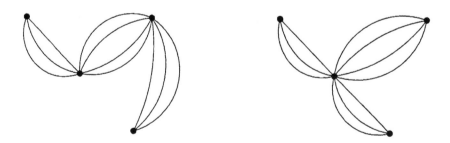

The duals are different: one has a vertex of order 10 and the other does not. Keep in mind that when you refer to the dual of a graph, you may need a diagram to describe which dual you mean.

2

Surfaces

2.1 The shape of the world. How do we know the world is round?

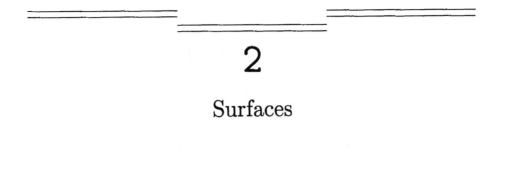

A sphere   A torus

Is it possible that we live on a torus, as opposed to a sphere? Pictures taken from space give uncontestable proof that the world is not a big donut, but people knew that the surface of the Earth was a sphere long before humans developed spaceflight. What was the evidence?

Task 2.1.1: List the evidence for why the surface of the Earth is a sphere.

Task 2.1.2: If the world was perpetually shrouded in dense fog and the terrain was extremely lumpy, would the evidence you gave in Task 2.1.1 still work? Do your methods permit the inhabitants to distinguish between the two worlds below?

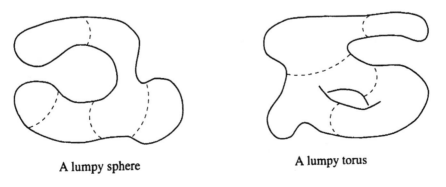

A lumpy sphere   A lumpy torus

If you confine yourself to just a small bit of your world then it is not possible to determine whether you live on a sphere, or a torus, or on some other surface. You must travel over the whole surface in order to truly understand it. The inhabitants of a world perpetually shrouded in dense fog would not have an easy task to determine the shape of their world. Keep that picture in mind as you read the rest of the chapter.

Note. In this chapter we are studying surfaces, not 'solid' objects. When we say *sphere* you should have in mind the outer skin of a ball. Similarly, the torus can be thought of as the outer skin of a donut.

One way to explore the whole world is to form a loop of people all facing the same way, and have them walk across the surface while holding hands.

A large group of people hold hands
to form a big loop on the Earth.

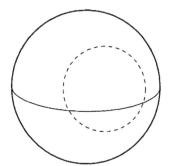

As the people walk forward they
all bunch together in one spot.

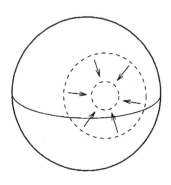

Task 2.1.3: Explain why any loop of people on a sphere will end up in one spot as they walk forward. Note: the loop may first get bigger before it gets smaller.

Definition. A loop on a surface is called **contractible** if the loop can be shrunk to a point without leaving the surface.

The result of Task 2.1.3 can be rephrased as, "On the sphere, all loops are contractible." The situation is quite different on the torus.

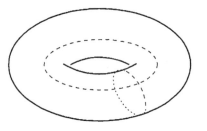

Task 2.1.4: Convince yourself that neither of these loops on the torus is contractible.

Keep in mind that the loops in Task 2.1.4 are on the *surface*. The loops must stay on the surface; they are not allowed to 'dig through the donut.'

Loops on a torus can have amazing properties unlike anything on a sphere:

A large loop of people are holding hands. They begin to walk forward.

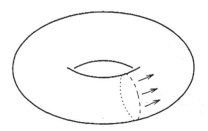

After a while they will have walked halfway around the world.

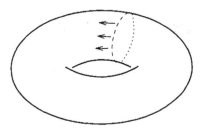

Eventually everybody gets back where they started.

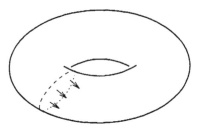

Task 2.1.5: Invent a story about the inhabitants of a torus planet and their attempts to determine the shape of their world. Would the members of the 'human chain' described above be surprised when they all returned to their original spot?

Task 2.1.6: What will happen
as this loop of people starts walking forward? Is this different
from what happens with the loop
shown above?

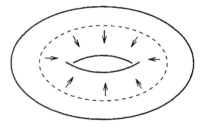

Task 2.1.7: How would your answers to the last two Tasks change if the loops
were on the lumpy torus shown at the beginning of the chapter, and everything
was shrouded in dense fog? If any of your answers involved measuring distances
or looking outside the torus, then those reasons are no longer valid.

Part of Tasks 2.1.6 and 2.1.7 asks you to determine how the inhabitants of
a torus can tell the difference between the two loops shown in Task 2.1.4. If
you are unable to measure distances properly, as would happen if the torus were
very lumpy, and if there was no way to look at the torus from the outside, then
it is *impossible* to distinguish between those two loops. This is an important
concept which may become more clear in the next section.

Task 2.1.8: The two loops in Task 2.1.4 cross at one point. Explain why, on
the plane, or on the sphere, it is impossible for two loops to cross at only one
point.

Task 2.1.9: Draw a loop on the torus which crosses each of the loops in
Task 2.1.4 exactly once.

Task 2.1.10: Draw a different pair of loops on the torus which cross at only
one point.

The book *The Shape of Space* [SoS], by Jeff Weeks, contains an amusing
story of a creature who travels around its world drawing lines on the ground.
Two of its longer trips result in a huge blue loop and a huge red loop drawn
around the world. It turns out that these loops cross each other only once. This
was quite confusing to the creature, but all it means is that its world is not a
sphere.

2.2 The flat torus

Here is a way to build a torus. Start with a square:

The arrows show you which
edges to glue together.

continued...

First glue the top edge
to the bottom to make
a cylinder.

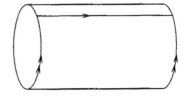

Then bend the cylinder...

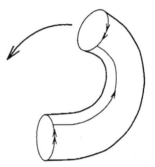

and glue the other edges to
complete the torus. The glue
lines become the loops shown
in Task 2.1.4.

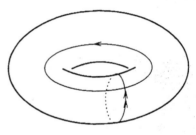

It is easy to draw a square, so we will use a square to represent a torus whenever possible. We call this the **flat torus**. Putting 'arrows' on the sides of a square shows that we mean for the opposite edges to be glued, with the understanding that the glue lines become the two curves shown directly above. Here is an example. Both figures represent the same loop on the torus:

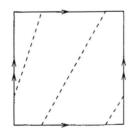

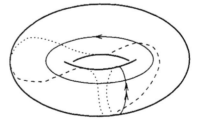

It is worth spending some time looking at that example. On the square, the places where the dotted line hits opposite edges must 'match up.' This ensures that, when the edges are glued, the dotted line will form one continuous loop. To verify that the two pictures represent the same thing, it is easiest to look at where the loop intersects the glue lines. The left picture shows the loop broken into three segments. You need only check that each segment in the right picture is drawn properly.

Here is another example. The segments are marked so that it is easier to see how things correspond.

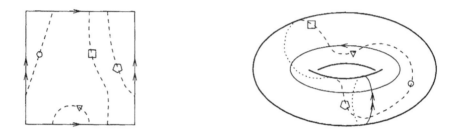

Task 2.2.1: Draw your answers to Tasks 2.1.9 and 2.1.10 on the flat torus.

The next section will provide plenty of practice drawing on the torus.

We mentioned earlier that it is not possible, in general, to distinguish between the two curves shown in Task 2.1.4. Thinking in terms of the flat torus will make this clear. If you repeat the procedure of gluing the square to give a torus, but first glue the left edge to the right edge to make a 'vertical' cylinder, the result will be a torus with glue lines the reverse of those shown above. Since it is impossible to distinguish between the edges of the flat torus, it is impossible to distinguish between the glue lines.

2.3 Graphs on the torus

In the previous chapter we showed that it is impossible to draw the graphs K_5 and $K_{3,3}$ without crossing edges. Actually, that statement isn't quite right. It is impossible to draw those graphs on the *plane* or the *sphere* without crossing edges. We will see that both K_5 and $K_{3,3}$ can be drawn on the torus. Here is one way to draw K_5, given in both representations:

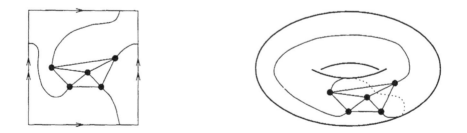

Usually it is easiest to first draw on the square, and then transfer everything to the other picture. When possible, show both pictures.

Task 2.3.1: Draw $K_{3,3}$ on the torus. Do the same for $K_{3,4}$ and $K_{4,4}$. There are several nice representations for $K_{4,4}$.

Task 2.3.2: Draw K_6 and K_7 on the torus.

The graphs K_8 and $K_{4,5}$ cannot be drawn on the torus. This is discussed in the next section.

Task 2.3.3: Draw the Petersen graph on the torus.

Task 2.3.4: Gluing opposite sides of a hexagon produces a torus. Use this representation to give a nice way to draw K_7.

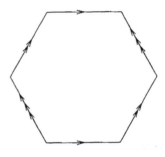

In Section 2.10 we will see why gluing opposite sides of a hexagon gives a torus.

2.4 Euler's formula, again

In the previous chapter we established Euler's formula $v - e + f = 2$ for any graph drawn on the plane. This formula also holds for any graph on the sphere.

Task 2.4.1: Explain why $v - e + f = 2$ holds for any connected graph drawn on the sphere.

Let's investigate $v - e + f$ for the torus. In these examples it is important to keep track of what is being glued together when counting edges and regions.

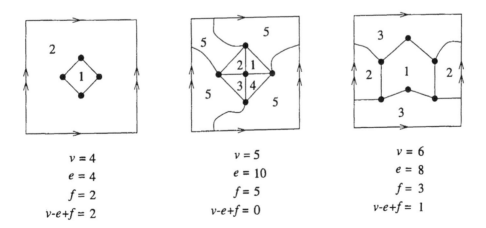

$v = 4$	$v = 5$	$v = 6$
$e = 4$	$e = 10$	$e = 8$
$f = 2$	$f = 5$	$f = 3$
$v-e+f = 2$	$v-e+f = 0$	$v-e+f = 1$

This looks like bad news. The value of $v - e + f$ doesn't appear to always be the same. Fortunately, the discrepancy is just an illusion. The key lies in the

separate regions of the graph. To get a workable formula, we must only use
graphs whose separate regions are *cells*. A region of a surface is called a **cell** if
all loops in that region are contractible.

This is region 2 of the first
graph given above. The loop
shown is not contractible, so
the region is not a cell.

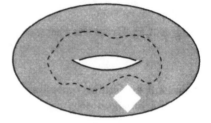

Here are more examples.

These are cells:

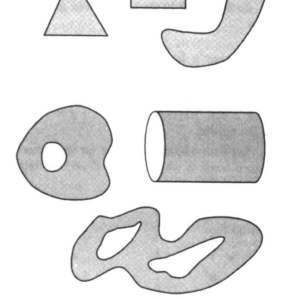

These are **not** cells:

Roughly speaking, a region is a cell if it doesn't have any 'holes' in it.

In the examples shown previously, we found $v - e + f = 0$ for the graph
which divides the torus into cells. This is Euler's formula for the torus.

Euler's formula for the torus. If a connected graph is drawn on the torus
so that the separate regions are cells, then $v - e + f = 0$.

The next two Tasks suggest methods of seeing why $v - e + f = 0$ is true for
the torus.

Task 2.4.2: Refer to Task 1.5.1. The graphs you drew in that Task can all be drawn on the torus, but the large outside region will not be a cell. It takes 2 more edges to turn that region into a cell. Putting in those edges adds 2 to e and C, and leaves all other quantities unchanged. Replace e by $e + 2$ and C by $C + 2$ in all of your formulas for the sphere, and you will wind up with $v - e + f = 0$ for the torus.

Task 2.4.3: A torus can be built from a sphere by cutting out two triangles and gluing the cut edges together:

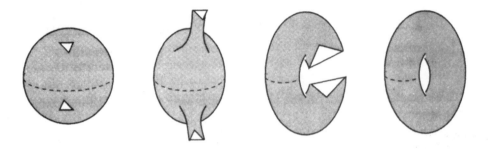

Altogether the triangles had 6 edges, 6 vertices, and 2 faces. The 2 faces were thrown away, and after the gluing, the two triangles became one triangle. So in going from a sphere to a torus we 'lost' 3 vertices, 3 edges, and 2 faces. Check that if you start with Euler's formula for the sphere, replace v by $v - 3$, replace e by $e - 3$, and replace f by $f - 2$, you get Euler's formula for the torus.

Task 2.4.4: What happens if you modify Task 2.4.3 by cutting out squares instead of triangles? Do you still get Euler's formula for the torus?

Task 2.4.5: Use $v - e + f = 0$ and $3f \leq 2e$ to show that K_8 cannot be drawn on the torus.

Task 2.4.6: Use $v - e + f = 0$ and $4f \leq 2e$ to show that $K_{4,5}$ cannot be drawn on the torus. Can $K_{3,6}$ be drawn on the torus? Either do it or explain why it is impossible.

Task 2.4.7: Suppose that removing one edge from a graph results in a graph which can be drawn on the sphere. Does it follow that the original graph can be drawn on the torus?

In Section 1.10 we discussed maps on the plane (or the sphere), and we mentioned that all planar maps can be 4–colored. One can also look at maps on the torus, and in this case the result is that all maps on the torus can be 7–colored. The proof is described in the Notes at the end of the chapter.

2.5 Regular graphs

In this section we study a special kind of graph.

Definition. A graph diagram is called **regular** if every vertex has the same order and every face is a cell of the same order.

We will often refer to regular graphs, although technically it is only correct to speak of regular graph diagrams. First we study regular graphs on the sphere.

Here are two regular graphs:

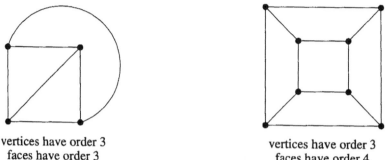

vertices have order 3   vertices have order 3
faces have order 3   faces have order 4

Those graphs are drawn on the plane, but it is easy to picture them drawn on the sphere.

Task 2.5.1: Explain why the dual of a regular graph diagram is also regular.

Our goal is to find all of the regular graphs on the sphere. There is an infinite list of uninteresting ones, three of which are:

Those graphs are commonly referred to as *cyclic graphs*.

The dual of a cyclic graph has multiple edges, so it is not actually a graph in our strict sense. You can decide for yourself if it should be included on our list of regular graphs.

Task 2.5.2: Find as many regular graphs as possible. Try to find at least three regular graphs in addition to the cyclic graphs and the two at the beginning of this section. For each one that you find, list the number of vertices, edges, and faces, and the orders of the vertices and the faces.

Now that you have a list of regular graphs, we want to check if the list is complete. Our secret weapon, as usual, will be Euler's formula $v - e + f = 2$.

Let M be the order of each vertex in a regular graph, and let J be the order of each face. First we will determine the possibilities for M and J. A little cleverness will save us lots of work. Each face of a graph must have order at least 3, so $J \geq 3$. By Task 2.5.1, the dual of a regular graph is regular. Taking duals switches vertices and faces, so taking duals switches J and M. Therefore, we also have $M \geq 3$.

In Section 1.11 we showed that any planar graph must have a vertex of order less than or equal to 5. Since all the vertices in a regular graph have the same order, we conclude that $M \leq 5$. By the same argument as above, we also have $J \leq 5$.

We have shown that J and M must both be between 3 and 5. It remains to be determined which possibilities can actually occur. You already listed some of these in Task 2.5.2.

Task 2.5.3: The cyclic graphs have $M = 2$. This contradicts $M \geq 3$. What is the problem?

To finish our analysis we must use Euler's formula $v - e + f = 2$. The important step here is to use the Counting Observation from Section 1.6: "If you add up the orders of all the vertices in a graph, the result equals twice the number of edges." In a regular graph all the vertices have order M, so we have $Mv = 2e$. Dividing by M gives $v = 2e/M$. We had a similar observation concerning faces: "If you add up the orders of all the faces in a graph, the result equals twice the number of edges." In a regular graph, this says $Jf = 2e$. Dividing by J gives $f = 2e/J$. Plugging our new formulas into Euler's formula $v - e + f = 2$ gives

$$\frac{2e}{M} - e + \frac{2e}{J} = 2.$$

For each of the possible values for M and J, we plug into that formula and solve for e. For example, if $J = 3$ and $M = 3$ then

$$\frac{2e}{3} - e + \frac{2e}{3} = 2,$$

which tells us $e = 6$. Sure enough, $J = 3$, $M = 3$, $e = 6$ describes one of the regular graphs we found earlier.

Another possibility is $J = 4$ and $M = 5$. This gives

$$\frac{2e}{5} - e + \frac{2e}{4} = 2,$$

which reduces to $e = -20$. That is nonsense, because the number of edges must be positive. We conclude that there is no regular graph with $J = 4$ and $M = 5$. The next Task finishes our study of regular graphs on the sphere.

Task 2.5.4: For each possible value of J and M, use the formula above to find the value of e. If the value of e is sensible, then find a regular graph with those values of J, M, and e. Note: there are 9 combinations of J and M to be checked. Two of them were already done in the text above.

The regular graphs on the sphere were first studied by the ancient Greeks in terms of the **regular solids**:

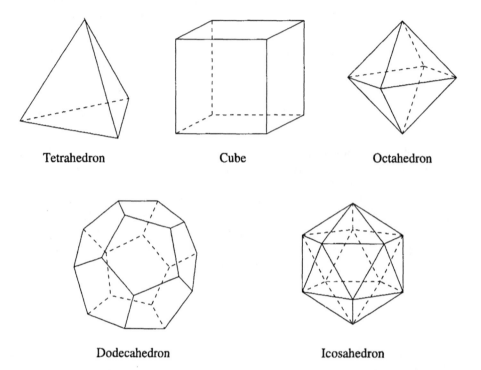

Tetrahedron   Cube   Octahedron

Dodecahedron   Icosahedron

For more information on the regular solids, see the book *Shapes, Space, and Symmetry* [SSS] by Holden.

Now we turn our attention to the torus. Part of our previous work can be reused: the relations $Mv = 2e$ and $Jf = 2e$ hold for a regular graph on any surface. Plugging into Euler's formula for the torus gives

$$\frac{2e}{M} - e + \frac{2e}{J} = 0.$$

We can factor the left side to get

$$e\left(\frac{2}{M} + \frac{2}{J} - 1\right) = 0.$$

We may assume $e \neq 0$, otherwise our graph would have no edges. So we must have

$$\frac{2}{M} + \frac{2}{J} - 1 = 0.$$

When solving that equation we must keep in mind the meaning of J and M. For example, $M = \frac{14}{3}$ and $J = \frac{7}{2}$ is a solution, but this is a nonsense solution because it is impossible for $\frac{14}{3}$ to be the order of a vertex. We must restrict ourselves to whole number values of J and M.

Task 2.5.5: Check that these are the only positive integer solutions to the above equation:

$$J = 3 \quad \text{and} \quad M = 6$$
$$J = 4 \quad \text{and} \quad M = 4$$
$$J = 6 \quad \text{and} \quad M = 3$$

Task 2.5.6: Draw regular graphs on the torus corresponding to each possibility in Task 2.5.5. You have already done some of these in Section 2.3.

2.6 More surfaces: holes

So far we have studied the sphere and the torus. Now we look at some other surfaces.

One way to get a new surface is to cut holes in a surface you already have.

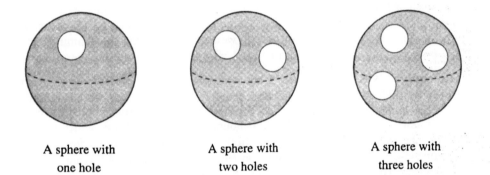

A sphere with   A sphere with   A sphere with
one hole   two holes   three holes

Surfaces can appear in disguised form. Each of these surfaces is 'the same' as the corresponding sphere with holes shown above:

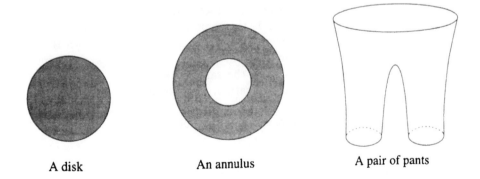

A disk   An annulus   A pair of pants

In a later section we will discuss what it means for two surfaces to be 'the same.' For now, it is sufficient to think of two surfaces as 'the same' if one can be smoothly deformed to give the other. For example, an annulus is 'the same' as a cylinder:

Task 2.6.1: Convince yourself that a disk is 'the same' as a sphere with one hole, an annulus is 'the same' as a sphere with two holes, and a pair of pants is 'the same' as a sphere with three holes.

Task 2.6.2: What common article of clothing is 'the same' as a sphere with four holes?

On 'quotes'. In the last page we have used the phrase 'the same' several times. Each time we were careful to put 'quote marks' around it. The quote marks are used to show that we are using words imprecisely. Specifically, since we have not given a precise definition of what it means for two surfaces to be *the same*, we put quote marks around it to show that we are aware of this shortcoming. This is a useful way of working with a concept which is not yet fully understood.

On equivalence. In Chapter 1 we encountered different–looking diagrams which we considered to be 'the same graph.' In this chapter we encountered different–looking surfaces which we consider to be 'the same.' In both cases there was the underlying idea that, although we had two things which weren't 'really' the same, we were going to think of them as 'the same' for our current purposes. This situation occurs so frequently that mathematicians have a special word to describe it. Instead of using saying 'the same,' we use the word **equivalent**. An example sentence is, "An annulus is equivalent to a sphere with two holes." Now, we still haven't given a precise definition of what *equivalent* means in this context, but at least we no longer have to use those annoying quote marks.

In everyday life the words "the same" rarely means "absolutely exactly the same." Take the phrase, "my car is the same as your car." That statement is perfectly sensible, and we know it doesn't mean that we both own the very same automobile! It could mean that our cars are the same make and model, or maybe it means that our cars are the same make, model, year, and color. If you heard that phrase in a conversation, you would be able to understand what it meant. In the same way, we use *equivalent* to mean "the same, as far as our current purposes are concerned." It is everybody's job to keep track of the current meaning of *equivalent*.

2.7 More surfaces: connected sums

We created new surfaces by cutting holes in a surface we already had. Another way to get a new surface is to 'glue together' two surfaces. The official name for this is **taking the connected sum** of the two surfaces. Here is an example where we take the connected sum of two tori:

Remove a disk from each
surface.

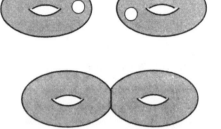

Glue the surfaces together
along the cut edges.

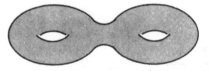

Smooth out the seam.

Surprisingly, nobody has invented a good name for the surface we just created. We will call it the **double torus**.

The above procedure can be applied to any pair of surfaces. Taking the connected sum of a torus and a double torus produces this surface:

We will call this the **triple torus**.

Task 2.7.1: Draw a sequence of pictures to show that if you take the connected sum of a surface with a sphere, the result is equivalent to the original surface.

Another way to prove the fact mentioned in Task 2.7.1 is: A sphere minus a disk is the same as a disk, so taking the connected sum with a sphere just replaces the disk which was removed from the original surface.

Combining our two methods of producing surfaces gives this catalog:
Spheres with holes:

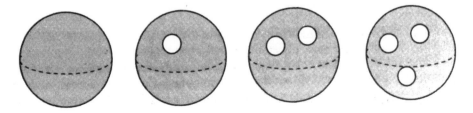

Tori with holes:

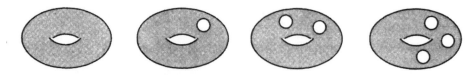

Double tori with holes:

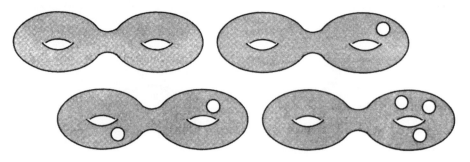

Triple tori with holes:

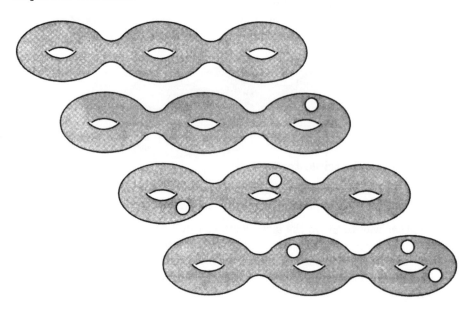

and so on....

Note that each of these surfaces can be described with just two numbers: the number of tori in the connected sum, and the number of holes.

We refer to the edges of the holes as the **boundary** of the surface. A sphere has no boundary, a torus with three holes has three boundary curves, and so on. To a creature living on the surface, the idea of a boundary curve is more natural than the idea of a hole. As the inhabitants move about on the surface they occasionally encounter an 'edge' to their world. From the outside we can see that they have walked up to the one of the holes, but on the surface it merely looks like the world ends. An inhabitant can walk along the edge until it gets back to where it started. To the creature, it just seems like there is a line on the ground which marks the boundary of the world. If the surface has more than one 'hole' then the inhabitant can walk along the surface to another boundary curve and walk around it until it gets back to where it started. By making marks on the ground it can count how many boundaries the world has. We can see that the creature is just counting holes, but our view from outside the world is unnatural. The natural perspective is that of the creature on the surface, and so boundary curves, not holes, are the preferred object of study. We will look at this further in a later section.

2.8 One–sided surfaces

Our list of surfaces is not complete. One surface we are missing is the *Möbius strip*, named after the German mathematician August Möbius. To make a Möbius strip, glue the ends of a strip of paper with a half–twist.

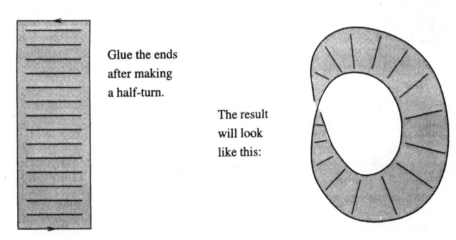

Glue the ends
after making
a half-turn.

The result
will look
like this:

Task 2.8.1: Imagine that you take a Möbius strip made of paper and you cut it in half down the center. What will be the result? What if you cut the resulting surface in half again? What if you cut the original Möbius strip into thirds? Make detailed predictions as to the result of each operation.

Task 2.8.2: Make some Möbius strips out of paper and check your predictions from the previous Task. If your predictions were wrong, try to figure out where you made an error.

The Möbius strip has two interesting properties: it has just one edge, and it has only one side. Check those properties with your paper model: you can trace the whole edge without picking up your finger, and you can go from 'one side' to 'the other side' by going around the strip. It is worthwhile to look back at the previous two Tasks and express your results in terms of the number of edges and sides of the surfaces.

Since the Möbius strip has only one side, it cannot be in the large catalog of surfaces we produced previously. If you look back on that list, you can easily convince yourself that all those surfaces have two sides.

A Möbius strip has one curve for its boundary. A disk also has one boundary curve. We can imagine gluing a disk onto a Möbius strip, producing a surface with no boundary. The resulting surface is called the **projective plane.**

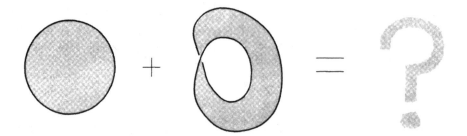

We put a fuzzy ? because it is impossible to accurately draw a projective plane. This should be believable, because you can imagine trying to sew a disk onto a Möbius strip: things will be easy at first, but no matter how stretchy the fabric is, you will get stuck before you are able to finish. However, our inability to draw the projective plane will not stop us from learning things about it.

The projective plane is the building block for all the one–sided surfaces. Just like we made more two–sided surfaces by taking the connected sum of tori, all one–sided surfaces are made by taking the connected sum of projective planes (and cutting holes in those surfaces). The connected sum of two projective planes is called the **Klein bottle**, named after Felix Klein. Since a projective plane minus a disk is a Möbius strip, the Klein bottle is equivalent to the surface you get when you glue together two Möbius strips. The reason it is called a 'bottle' is shown in the Project on one–sided surfaces, at the end of the book.

The projective plane and the Klein bottle have useful representations as squares with opposite sides glued:

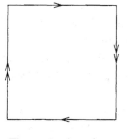

The projective plane

The Klein bottle

Note that the arrows do not all face the same way, sometimes we 'make a twist' when gluing. These representations are useful for drawing graphs on the surfaces:

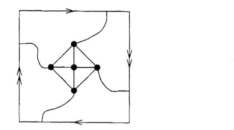   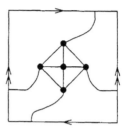

K$_5$ on the projective plane   K$_5$ on the Klein bottle

It is important that edges of the graph match up properly after we glue the sides of the square. The presence of a 'twist' makes things a bit more tricky than when we were dealing with the torus. Check that the above examples are drawn correctly.

This ends our discussion of one–sided surfaces. To learn more about these interesting surfaces see the Project on one–sided surfaces at the end of the book. In the next section we return to two–sided surfaces.

2.9 Identifying two–sided surfaces

In Section 2.7 we gave a complete list of two sided surfaces. There remains the problem: given a two–sided surface, how do we decide which surface it is? This is not as trivial as it sounds. The surfaces below are two–sided, but it is not obvious where each fits on our previous list.

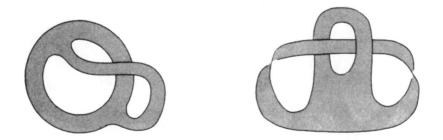

Before identifying these surfaces in terms of our previous list, we must develop our concept of what it means for two surfaces to be equivalent. Recall that we take the perspective of someone living on the surface. If you are confined to the surface then you can't distinguish many things which are evident from the 'outside.' For example, suppose you live on an annulus. You would be oblivious to the following procedure:

Choose any line going from
one boundary to the other.

Cut along the line and twist
the surface in various ways.

Glue the cut edges together
the way they were originally.

The cut edge was put back exactly the way it started, so if you are confined to
the surface it is impossible to distinguish between the original annulus and the
twisted annulus.

Task 2.9.1: Show that each of these is equivalent to an annulus.

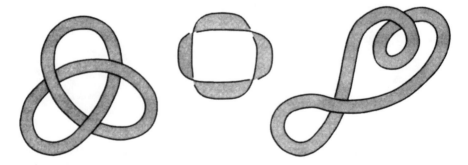

Also, the figure on the cover of this book is equivalent to an annulus.

It is difficult to write down a useful and completely rigorous definition of
what it means for two surfaces to be equivalent. For our purposes it is sufficient

to think of it as "indistinguishable by people who are confined to the surface, have poor eyesight, and are bad at measuring distances." Two important ways of manipulating a surface to get an equivalent surface are: bend and stretch without tearing it; and cut it apart and twist it around, reassembling the pieces so they fit together the same as originally.

On topology. The above discussion of a mystery world inhabited by people who have poor eyesight and are bad at measuring distances might remind you of the insect world from Chapter 1. This is a main theme of this book: we ignore size and distance, and just look at how things are connected. For example, when we studied graphs the only thing we cared about was which vertices were connected to which other vertices. And as we study surfaces, we only care how the surface is 'connected to itself,' the outside appearance of the surface being of only minor consequence. Mathematicians call this area of mathematics **topology**. Note: do not confuse topology, a branch of mathematics, with topography, the study of mapmaking.

Now we develop an organized way to distinguish surfaces. The surfaces on our list can each be described as the connected sum of some number of tori, each with some number of holes. Counting holes is the easier part. Each hole has one boundary curve, so we count holes by counting boundary curves. To do this, start at one point on the edge of the surface, and trace that edge until you end up back where you started. If there is a part of the boundary which hasn't been traced, then start over on an untraced part. Keep going until you trace the entire edge. The number of times you started tracing is equal to the number of boundary curves. An easy way to keep track is to trace each boundary curve with a different color pen.

Task 2.9.2: Check that these are labeled correctly:

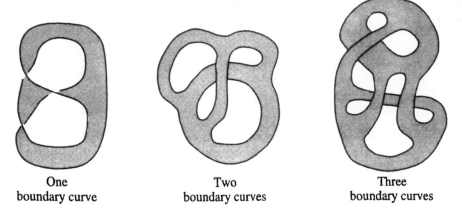

| One boundary curve | Two boundary curves | Three boundary curves |

Task 2.9.3: Count the number of boundary curves on the two surfaces at the beginning of this section.

If a two-sided surface has no boundary then it is a sphere, or a torus, or a double–torus, or.... If a surface has one boundary curve then it is a *something* with one hole. The problem is to determine the *something*. The key to this is our friend $v - e + f$. In the case of the sphere we found $v - e + f = 2$ and for the torus we found $v - e + f = 0$, so we can use $v - e + f$ to distinguish between the sphere and the torus. This suggests that it will be useful to find $v - e + f$ for our other surfaces.

Since f stands for the number of separate regions...

Task 2.9.4: Explain why cutting a hole in a surface decreases $v - e + f$ by 1.

So, a sphere with one hole has $v - e + f = 1$, a sphere with 5 holes has $v - e + f = -3$, a torus with 2 holes has $v - e + f = -2$, and so on.

If we knew $v - e + f$ for the double torus, triple torus, and so on, then we could determine $v - e + f$ for all the 2–sided surfaces.

Task 2.9.5: Explain why taking the connected sum of a surface with a torus decreases $v - e + f$ by 2. So, for the double torus $v - e + f = -2$, for the triple torus $v - e + f = -4$, etc. The reasoning in Task 2.4.3 may be helpful here.

Task 2.9.6: Make a chart of all the 2–sided surfaces and write the value of $v - e + f$ next to each one.

Now we have an infallible way to identify any 2–sided surface: first count the number of boundary curves, then find $v - e + f$ and look up the answer in your chart from Task 2.9.6. To find $v - e + f$, draw a graph on the surface which divides the surface into cells. The usual method is to decide how many edges need to cut across the surface to divide it into cells, put vertices at the ends of those edges, and connect the vertices by putting graph edges along the entire boundary of the surface. Here are two examples:

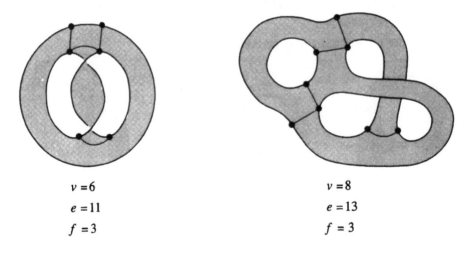

$v = 6$ $v = 8$

$e = 11$ $e = 13$

$f = 3$ $f = 3$

The first surface has three boundary curves and $v - e + f = -1$, so it is a sphere with three holes. The second has two boundary curves and $v - e + f = -2$, so it is a torus with two holes. In both examples we used more vertices and edges than were absolutely necessary to cut the surface into cells. Adding too many vertices and edges makes it difficult to correctly count v, e, and f, but if you fail to cut the surface into cells then you won't get the right answer.

Task 2.9.7: Identify all the surfaces which have appeared in this section.

2.10 Cell complexes

This section finishes our study of two–sided surfaces.

Below are assembly instructions for building a surface. Edges labeled with the same number get glued together, and the arrows show which way to match the edges.

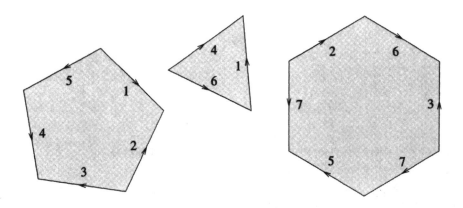

This is an example of a **cell complex**: a bunch of cells which are glued together to make a surface. We first encountered this idea back when we began studying the flat torus. Gluing the sides of a square to get a torus is an example of a cell complex with just one cell.

Task 2.10.1: Is it possible to make the above surface out of actual pieces of paper?

We use our usual method to determine what surface is formed from the cell complex. First we count boundary curves. In this example, every edge gets glued, so the surface has no boundary, so the surface is a *something* with no holes. Next we find $v - e + f$. The glue lines form the edges of a graph, so we will use that graph to find v, e, and f. The cell complex has 3 cells, so $f = 3$. There are 7 glue lines, labeled 1 through 7, so $e = 7$. The corners of the cells will become vertices, but many corners will be glued together at the same vertex. When a pair of edges get glued, that also glues together the corners at the respective ends of the edges. To count vertices, label a corner and then trace through all the glued edges and give the same label to all the corners which get glued to the original corner. If some corner hasn't been labeled, then repeat the process. The number of labels you need is equal to the number of vertices. Here is the result for the above example:

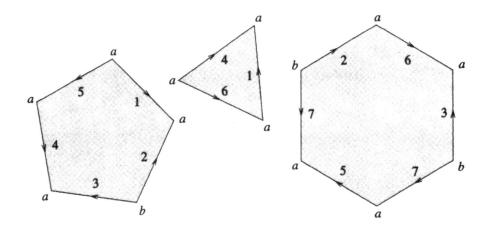

There are two different vertices: a and b, so $v = 2$. Therefore $v - e + f = -2$, so the surface is a double torus.

Task 2.10.2: Identify this surface:

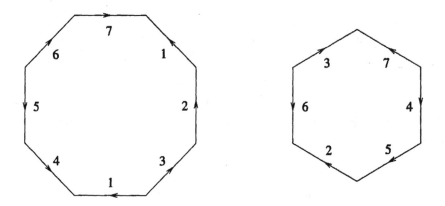

Task 2.10.3: In Task 2.3.4 we stated that gluing opposite sides of a hexagon gives a torus. Verify that this is true.

Task 2.10.4: What surface do you get when you glue opposite sides of an octagon? decagon? dodecagon? (Those polygons have 8, 10, and 12 sides, respectively.)

We have repeatedly stated that our list of two-sided surfaces is complete. The easiest proof of this fact uses cell complexes. This is briefly discussed in the Notes at the end of the Chapter.

One type of cell complex is of particular importance:

Definition. A **triangulation** is a cell complex where:

– All the cells are triangles.

– Two different triangles in the cell complex share either exactly one vertex, or exactly one edge (and the vertices at the ends of that edge), or else they don't meet at all.

– No triangle shares a vertex or edge with itself.

The triangles can look bent or twisted, but they must still be cells whose border has three vertices and three edges.

Here are some of the ways that triangles are *not* allowed to border in a triangulation:

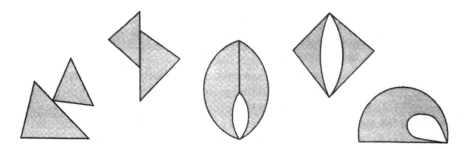

Task 2.10.5: Explain why the number of triangles in a triangulation must be even. Hint: this is related to Task 1.6.5.

The answers to the next Task have already appeared in this Chapter.

Task 2.10.6: Find a triangulation of the sphere using 4 triangles. Find a triangulation of the torus using 14 triangles.

It is impossible to triangulate the sphere with fewer than 4 triangles. We must use an even number of triangles, so the only possibility would be to use 2 triangles. You can easily make a sphere by gluing together two triangles, but the result is not a triangulation.

It is also impossible to triangulate the torus with fewer than 14 triangles. You can show this by combining $v - e + f = 0$, the idea behind Task 2.10.5, and the observation that a graph with n vertices has at most as many edges as K_n. There is a simple way to triangulate the torus with 18 triangles: cut the flat torus into 9 squares, then divide each square into two triangles.

2.11 Notes

Note 2.11.a: The statement "The Earth is a sphere" is inaccurate in several ways. It is more precise to say "The surface of the Earth is a sphere." That statement is also inaccurate because certain geographic features, such as arches and tunnels, cause the surface of the Earth to be quite complicated. It would be impossible to determine exactly which two-sided surface the surface of the Earth is, because the answer depends on how fine of a scale you use to measure.

If you measure things on a very large scale then the surface of the Earth is a good approximation to a sphere. Outside of pure mathematics, that is the best you could hope for.

Note 2.11.b: The plural of *torus* is **tori**.

Note 2.11.c: Although the word *equivalent* will have different meanings in different contexts, in each case we require the following:

 - Everything is equivalent to itself.

 - If A is equivalent to B, then B is equivalent to A.

 - If A is equivalent to B, and B is equivalent to C, then A is equivalent to C.

Note 2.11.d: It is traditional to use the letter F to represent a surface. This is because the German word for "surface" is "fläche."

Note 2.11.e: For any surface F, the number $v - e + f$ is called the **Euler characteristic** of the surface, and it is denoted by $\chi(F)$. For example, if T is the torus, then we would write $\chi(T) = 0$. Note: χ is the Greek letter chi, pronounced 'kī,' as in 'kite.'

Note 2.11.f: The two–sided surfaces with no boundary are all connected sums of tori. The number of tori is called the **genus** of the surface, and it is commonly denoted by g. The connection between genus and Euler characteristic is $\chi(F) = 2 - 2g$. If the surface also has b boundary curves, then $\chi(F) = 2 - 2g - b$.

Note 2.11.g: The distinction between two–sided and one–sided surfaces may seem natural, but this actually reflects our prejudice as 3–dimensional beings. First, you should not think of the inhabitants of a surface as people walking on top it, but rather as 2–dimensional beings embedded in it. These beings do not think in terms of 'sides' of the surface. They can only conceive of two directions of motion, so the possibility of a third direction of motion towards the 'top' is beyond their comprehension. Fortunately, they can still understand our distinction between one– and two–sided surfaces by using the related concept of orientability. We will see that the two–sided surfaces are orientable and the one–sided surfaces are nonorientable.

It is easiest to explain by an example. The following illustrates that the Möbius strip is nonorientable. Keep in mind that the objects 'on' the surface are actually 2–dimensional objects 'in' the surface.

Take a right–hand glove and slide it around a Möbius strip.

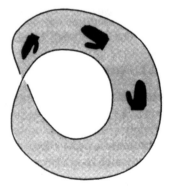

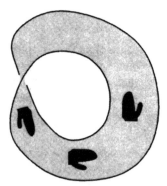

Keep sliding until it returns to
its original place. It is now a
left–hand glove!

A surface is **nonorientable** if it is possible to turn any shape into its mirror–
image by moving it along some path in the surface. Such a path is called an
orientation reversing path. A surface is **orientable** if it isn't nonorientable.
You should convince yourself that the two–sided surfaces are orientable, meaning
that it is impossible to turn a right–hand glove into a left–hand glove by moving
around on those surfaces.

An alternate definition of orientability uses cell complexes. To **orient a
cell** means to specify a direction of travel around its boundary.

An oriented cell.

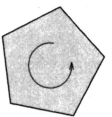

A cell complex is **orientable** if we can orient each cell, such that at the edge
where two cells meet, the orientations from each of the two cells are in opposite
directions. We say that a surface is orientable if it is given by a cell complex that
is orientable. This is equivalent to the previous definition of orientable. Note:
it is best to think of an orientation of a cell not as a cyclic ordering of its edges,
but rather as a cyclic ordering of its vertices. In two dimensions the distinction
is irrelevant, but in higher dimensions the latter view is better.

Note 2.11.h: Here is one way to prove that our list of two–sided surfaces is
complete:

- Show that all surfaces can be written as a cell complex.

- Show that the cell complex can be chosen to have just one cell.

- Show that the cell complex can be chosen to have just one cell and one
  vertex.

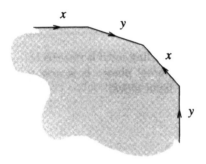

- Show that the complex can be chosen to have just one cell with the sides glued in groups of four like so:

- Finally, conclude that the cell complex gives one of the surfaces on our list. The representation of a surface as one cell with edges glued in groups of four (as above) is called the **standard form** for the surface.

Standard form of the double torus:

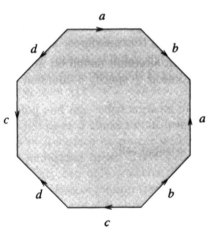

Note 2.11.i: On the plane, or the sphere, any map can be 4–colored, and some maps cannot be done with fewer than 4 colors. The proof of this is very difficult. On the torus, any map can be 7–colored, and some maps require 7 colors. This can be proven by an easy modification of the 6–coloring method from Chapter 1. Here are the steps:

- Use $v - e + f = 0$ to show that any graph on the torus has a vertex of order 6 or less.

- Modify the 6–coloring method of Chapter 1 to show that any graph on the torus can be 7–colored.

- Give an example of a graph on the torus which requires 7 colors.

The first step requires a small bit of calculation, the second is nearly identical to the planar method of Chapter 1, and the third has already been done in Task 2.3.2.

It is interesting that the Four–Color Theorem for the sphere is very difficult, while the Seven–Color Theorem for the torus is very easy.

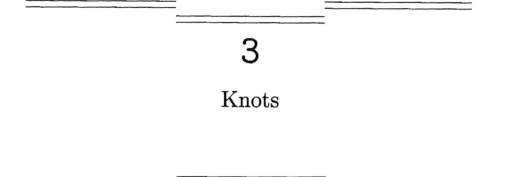

3

Knots

3.1 The view from the outside. In the previous two chapters we were concerned with graphs and surfaces as viewed from the object itself. Now we shift our perspective and consider objects as viewed from the outside.

Here are three knots. The pictures are meant to represent a piece of string which has been twisted around and then had its ends sealed together.

If you were put on the knot itself, all you could tell was that it is one continuous closed loop, so a knot isn't interesting unless you view it from the outside. The manipulations we do to mathematical knots are exactly the same as those which can be done to an actual piece of string. For example, the following manipulation takes the third knot above and rearranges it to give the first knot.

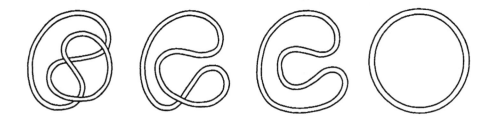

Visualizing and drawing knots can be difficult at first, so you may find it helpful to keep on hand an old shoelace or other piece of string to tie into the knots we study.

The picture we draw of a knot is called a **knot diagram**, and there are

many different diagrams representing the same knot. Think of the knot as
floating around in space. Each diagram of that knot is one of the many pictures
we might see as we look at the knot from different perspectives. Two knot
diagrams are called **equivalent** if one can be manipulated to look like the other.
Another way to put it is, two knot diagrams are equivalent if they both represent
the same knot. The distinction between a knot and a knot diagram is similar
to the distinction between a graph and a graph diagram from Chapter 1. Also,
the concept of two knot diagrams being equivalent is similar to the idea of two
surfaces being equivalent from Chapter 2. It may be helpful to look back at the
discussion of *equivalence* in Section 2.6.

From now on we will represent knot diagrams by drawing a solid line and
leaving a 'gap' to indicate the under–crossing:

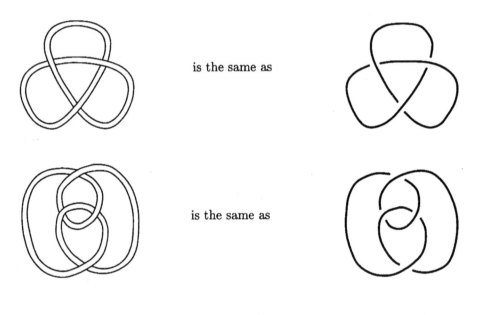

is the same as

is the same as

The individual segments that we draw in the diagram are called the **strands** of
the knot diagram. The first diagram above has three strands, and the second
has four strands. The place where we leave a 'gap' to show that one part of the
knot passes 'above' another part is called a **crossing** in the diagram.

Task 3.1.1: What is the relationship between the number of crossings and the
number of strands in a knot diagram? Explain why this relationship holds.

3.2 Manipulating knots

There are lots of ways to manipulate knots. The best way to show your
manipulations is as a sequence of diagrams, where a normal person can easily
see exactly what you did at each step. For example, here is one possible step in

a manipulation.

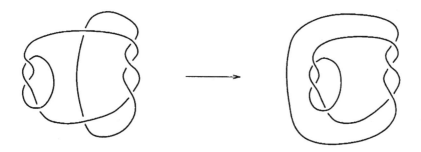

The diagrams are intended to show that we took one strand of the knot and dragged it underneath and to the left. If each step of your manipulation involves moving just one strand, then you can be fairly sure that other people will be able to follow what you are doing. Here is another example of moving just one strand.

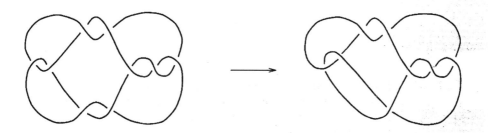

Some moves can't be easily shown 'one strand at a time.' For example:

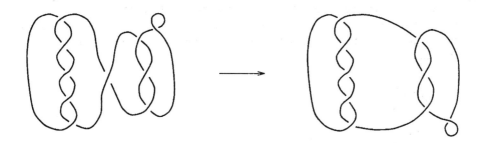

The diagrams should convey the idea that we took one part of the knot and 'flipped it around.'

All the manipulations you do to a diagram result in a different diagram of the same knot. Some of the diagrams will look incredibly complicated, while others may look simple. It is natural to try and manipulate a knot diagram to make it look as simple as possible. How simple it can get depends on which knot it represents.

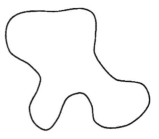

The simplest knot is called the
unknot. It is the only knot
which has a diagram with no
crossings.

Note that we said the unknot *has* a diagram with no crossings. A diagram with
crossings could still possibly be the unknot. For example, two of these diagrams
represent the unknot:

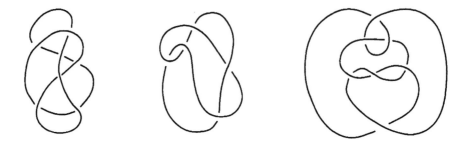

Task 3.2.1: Rearrange each of the above diagrams so that it has as few crossings
as possible. Two of them are the unknot, so they can be rearranged to have no
crossings. The other one isn't the unknot, so no matter how hard you try, you
can't get rid of all the crossings.

Task 3.2.2: Make up some mathematical puns using "un," "not," "knot," and
"unknot." Explain them to your friends.

Some of our goals will be to classify knots, find interesting properties of
knots, and devise clever ways to determine if two diagrams represent the same
knot. But first we need some hands–on experience. Here are a few simple knot
diagrams:

Later we will discuss that each of those diagrams represents a different knot,
and each is drawn with *as* few crossings as possible.

Task 3.2.3: For each of the following diagrams, try to determine if it represents one of the knots shown above.

As you may have noticed, it is not always easy to tell when two knot diagrams are equivalent. Showing that two knots are different can be even more difficult because you must show that no manipulation, no matter how clever or complicated, can rearrange one to look like the other. These are some of the things we will address later in the chapter.

3.3 Lots of knots

In this section we discuss the problem of creating a 'catalog' of knots.

In Chapter 1 we made a list of some graphs, arranging them by the number of vertices and edges. In Chapter 2 we made a list of surfaces, arranging them by the number of holes and the number of tori in the connected sum. If we want to make a list of knots, then we need some way to organize the list. Surprisingly(?), nobody has found a good way to write down a list of all knots. You will see that the method we will use has its shortcomings, but until someone invents a better way, it is the best choice we have. The method is based on the idea of *crossing number*.

The smallest number of crossings needed to draw a knot is called the **crossing number** of the knot. A great shortcoming in the definition of crossing number is that each knot has many different diagrams. Given a knot diagram, it might take a lot of work to reduce it down to the smallest possible number of crossings. It can even be difficult to determine when a knot has been drawn with the smallest possible number of crossings.

Recall that the unknot can be drawn with no crossings. In other words, the unknot has crossing number 0. If you look back at the examples in the previous section, you will find knots with crossing numbers 3, 4, 5, 6, and 7.

Task 3.3.1: Draw a few knot diagrams with 1 or 2 crossings. What knot does each of your diagrams represent?

The next few Tasks use the idea of a *projection* to show that there are no knots with crossing number 1 or 2, and to find all knots with crossing number 3.

A **knot projection** is like a knot diagram, except that the over/under crossings are not shown. The name comes from the idea that it looks like the 2–dimensional shadow of a 3–dimensional knot. Here are some examples:

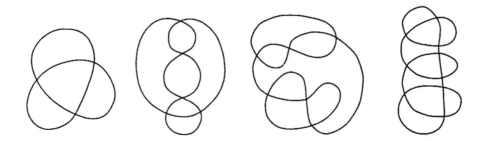

Given a knot projection, you can turn it into a knot diagram by choosing over/under at each crossing. There is a choice at each crossing, so there can be many different knots with the same projection.

Task 3.3.2: Find all projections with one crossing (there are only two of them). Show that every way of turning those projections into a knot diagram results in the unknot. Thus there are no knots with crossing number 1.

Task 3.3.3: Find all projections with two crossings (there are between 5 and 10 of them). Show that every way of turning those projections into a knot diagram results in the unknot. Thus there are no knots with crossing number 2.

Task 3.3.4: Find all projections with three crossings (there are between 20 and 25 of them). Determine which of those projections represent the unknot no matter how the projection is turned into a diagram. How many different knots can be obtained from the remaining diagrams?

Task 3.3.5: Begin making a list of projections with 10 crossings. Write down a dozen or so of them and then describe how you would go about systematically completing the list (assuming you had enough time to do it).

Task 3.3.6: If you draw any continuous curve which crosses itself a bunch of times and ends up back where it started, the result is a knot projection. See Task 1.10.7 for an example. Now modify the procedure so that when you are about to cross a line which you drew previously, you pick up your pen and then put it down again on the other side. Keep drawing as much as you want, but always pick up your pen as you cross a previously drawn line, and eventually finish at the same point where you started. The result will be a knot diagram. Draw a few examples and see what knot the diagram represents.

For more than 100 years work has been spent classifying and cataloging knots. There is no reason for us to repeat all that work, so you can turn to the end of the chapter to see a chart of all of the knots with 7 or fewer crossings. There are 26 of them. Spend some time looking at the similarities and differences among the various knots. After the unknot, the next three simplest knots also have common names.

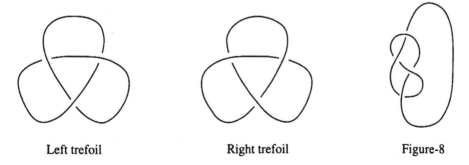

Left trefoil   Right trefoil   Figure-8

Other diagrams of those knots appeared in Task 3.2.3.

The names "left" and "right" given to the two forms of the trefoil might seem arbitrary, but they actually have a perfectly logical basis. To understand this, we need the concept of an *oriented knot*. An **orientation** of a knot is a choice of direction of flow around the knot. Here are two oriented knots.

Given a knot, you can orient it by drawing one 'arrow,' and then following that arrow around the knot drawing another arrow on each strand. The choice of direction of the first arrow is arbitrary, but after that there will be only one way to draw the other arrows to maintain the same direction of 'flow.'

Each crossing in a knot diagram can be classified as either 'left–handed' or 'right–handed.'

A left-handed crossing

A right-handed crossing

Here is how the names are determined. Take your right hand and point the thumb in the direction of one of the strands in the 'right–hand' crossing. Curl your fingers as if you were grabbing that strand, and your fingers will point in the direction of the arrows on the other strand. If you use your left hand on a right–hand crossing, then it won't work and your fingers will point the wrong way. In the two knots shown above, the one on the left has four right–handed and two left–handed crossings, while the one on the right has two left–handed and two right–handed crossings.

Task 3.3.7: Orient a right trefoil knot and a left trefoil knot. Are the names 'right trefoil' and 'left trefoil' appropriate?

Task 3.3.8: Orient your favorite knot diagram and write down which crossings are right–handed and which are left–handed. Then reverse the orientation of the knot (that is, reverse all the arrows). How have the crossings changed?

On the chart you will find some pairs of knots with the same name. The two knots of the pair have the same projection, but all the crossings in the diagram are 'opposite.' We say that the diagrams are **mirror images** of each other. Sometimes a diagram and its mirror image represent different knots, such as the left and right trefoil. And sometimes a diagram and its mirror image represent the same knot, as in the next two Tasks.

Task 3.3.9: Show that these diagrams represent the same knot.

Task 3.3.10: Show that these diagrams represent the same knot.

The last two tasks illustrate the difficulty of finding a 'good' diagram of a knot. The diagrams in Task 3.3.9 make it easy to show that the figure–8 knot is its own mirror image. The diagrams in Task 3.3.10 show that the figure–8 knot has a nice symmetry. Both sets of diagrams are interesting, and which one you consider to be 'better' will depend on what purpose you have in mind.

Task 3.3.11: Hold this diagram up to a mirror and draw the image. How does the knot you drew compare to what we have called the 'mirror image' of a knot?

The chart of knots at the end of the chapter shows both a knot and its mirror image (when those are different knots). Knot tables in other books save space by only showing one of a mirror image pair.

We said that the chart has all the knots with up to 7 crossings. That was a lie. Neither of these knots appears on the chart:

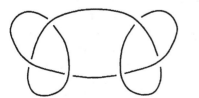

Both of those knots are an example of a *connected sum* of knots. Here is the procedure for making the connected sum of two knots.

Start with any two knots.

Remove a small segment from each knot.

Connect the loose ends *without introducing any new crossings.*

The result is called a **connected sum** of the two original knots. We used similar terminology, and a similar procedure, in the chapter on surfaces.

A knot which can be written as a connected sum of two nontrivial knots is called a **composite knot**.

Task 3.3.12: Two composite knots appeared in Section 3.2, and one appeared in Task 3.2.3. For each one, find two knots whose connected sum makes up the composite knot. Do the same for the two composite knots shown on the previous page.

A knot which cannot be written as a connected sum of two nontrivial knots is called a **prime** knot. *The chart at the end of the chapter only shows prime knots.* Given a composite knot, it can be written as the connected sum of two knots. If those two knots are not prime then each can also be written as another connected sum. The process continues and the hope is that eventually one gets down to prime knots, and those can be identified using a chart like the one at the end of the chapter. For example, the following knot is the connected sum of four prime knots, and those four knots can be found on the chart.

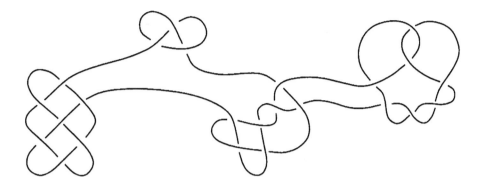

It is a fact that a composite knot can always be written as a connected sum of prime knots, but the proof is too complicated for us to address in this book.

There are several subtle questions we haven't addressed. For example, can you take the connected sum of two knots and get the unknot? Can a composite knot be written as a connected sum of prime knots in more than one way? Is there more than one way to take the connected sum of two knots? One of these questions is addressed by the next Task.

Task 3.3.13: Show that these diagrams represent the same knot.

3.4 Alternating knots

Put your finger on a knot diagram and begin tracing it. At each crossing, the strand you are on will either go 'over' or 'under' the other strand. If the pattern of crossings you encounter goes over–under–over–under..., continually switching back and forth from over to under, then we say the diagram is **alternating**. An **alternating knot** is a knot that has an alternating diagram.

You can check that all of the knots in the chart at the end of the chapter are alternating. This might lead you to guess that every knot has an alternating diagram. That guess would be wrong. Here is an example.

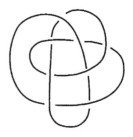

It is impossible to rearrange
this knot to get an alternating
diagram.

Task 3.4.1: In the previous section is shown a connected sum of a left trefoil
and a right trefoil. Rearrange that knot to get an alternating diagram.

Task 3.4.2: Spend a few minutes trying to find an alternating diagram of the
knot shown above. You will not succeed, because that knot isn't alternating,
but you should make a few attempts anyway.

One reason alternating diagrams are important is this useful fact.

Almost True Fact. An alternating diagram cannot be rearranged to have
fewer crossings.

We say 'almost true' because there is one way in which the fact can fail,
and it is easy to detect when this one exception occurs. We have encountered
such exceptions earlier in the chapter.

Task 3.4.3: Find alternating diagrams which violate the Almost True Fact.
Determine what they have in common, and find a concise way to describe the
exceptions to the ATF.

Task 3.4.4: Draw a knot diagram with 10 crossings which cannot be simplified
to have fewer crossings. Then do one with 20 crossings.

Task 3.4.5: Is it always possible to turn a knot projection into an alternating
knot diagram? If 'yes,' explain why. If 'no,' give an example of a projection
which can't be turned into an alternating diagram.

Task 3.4.6: Is it always possible to turn a knot projection into a diagram of
the unknot? If 'yes,' explain why. If 'no,' give an example. Hint: Task 3.3.6.

Task 3.4.7: Draw some diagrams where the pattern of crossings goes o–o–u–
u–o–o–u–u..., where o=over and u=under. Which knots can be represented by
diagrams of this type? Can all of those diagrams be simplified to have fewer
crossings?

3.5 Unknotting number

Suppose we do an 'illegal' move of switching one crossing of a knot.

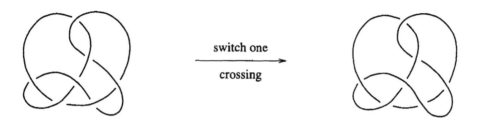

switch one

crossing

Task 3.5.1: Determine which knot appears on the right side above. That is, identify it as one of the knots shown in the chart at the end of the chapter.

Task 3.5.2: Determine what other knots occur when different crossings are switched in the diagram on the left side above. Can you obtain the unknot by switching one crossing?

Task 3.5.3: Determine which of the knots with 6 or fewer crossings can be turned into the unknot by changing one crossing.

Task 3.5.4: Does switching one crossing in a diagram always result in a diagram of a different knot?

The **unknotting number** of a knot is the least number of crossing changes which are needed to turn the knot into the unknot. We have seen that the trefoil and figure–8 knot have unknotting number 1. The 5_1 knot has unknotting number 2.

Task 3.5.5: Is it reasonable to think of a knot with a large unknotting number as being 'more knotted' than a knot with a small unknotting number?

Warning. The unknotting number is defined in terms of the *knot*, not the *diagram*. If it takes a certain number of crossing changes to turn one diagram into the unknot, there could possibly be another diagram of the same knot which requires fewer crossing changes to give the unknot. An example of this is given in the Notes at the end of the chapter.

A knot with a diagram of this form is called a **twist knot**.

The diagram can have any number of 'twists' in its lower portion. On the chart, the 5_2, 6_1, and 7_2 knots are drawn in a way which makes it clear that they are twist knots.

Task 3.5.6: Draw diagrams of the trefoil and figure–8 which make it clear that they are twist knots.

Task 3.5.7: Show that this
is a twist knot.

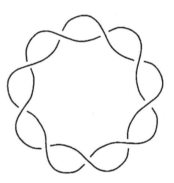

Task 3.5.8: What is the unknotting number of a twist knot?

Task 3.5.9: The 3_1, 5_1, and
7_1 knots are the beginning of a
family of knots, the next mem-
ber of which is the 9_1 knot:

Next comes the 11_1 knot, the 13_1 knot, and so on. Determine the unknotting
number of each of those knots. (Do the first few and then make a guess.)

Note: it is not feasible for you to be absolutely sure that your answer to
Task 3.5.9 is correct. This is because it is difficult to rule out the possibility
that the knots could be rearranged to give a diagram which can be unknotted
in fewer crossings. However, mathematicians have shown *in this particular case*
that no such rearrangement is possible, so your answer to Task 3.5.9 is probably
correct.

A **family of knots** is an informal term used to describe a list of knots where
each successive knot is obtained from the previous one by a simple process. The
twist knots are an example, as are the knots 3_1, 5_1, 7_1, ...

Task 3.5.10: Invent your own family of knots.

Task 3.5.11: Given a knot diagram, can you always make crossing changes in
that particular diagram, without doing any rearranging, to get a diagram of the
unknot? Hint: Task 3.4.6.

Task 3.5.12: What relationships are there between crossing number and unknotting number? For example, if a knot has crossing number 15, how large or
how small can its unknotting number be?

3.6 Links

Here are some links.

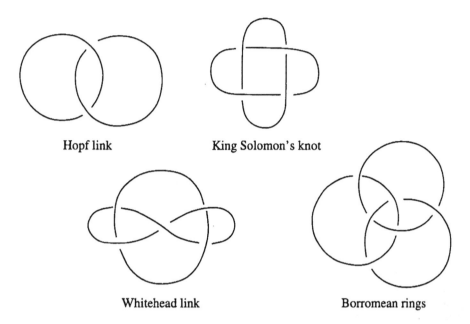

Hopf link   King Solomon's knot

Whitehead link   Borromean rings

A link is like a knot, except that it is made from more than one 'loop.' The separate 'loops' of the link are called the **components** of the link. In the examples above, the Borromean rings have three components, and the other links have two components. King Solomon's knot has a misleading name because it is a link, not a knot. Apparently it wasn't named by a mathematician!

The manipulations we do to a link are the same as the manipulations we do to a knot.

Task 3.6.1: For each diagram, determine if it represents one of the links shown above.

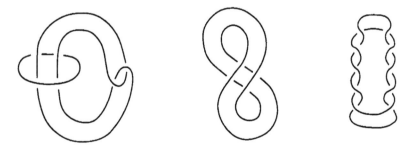

Task 3.6.2: We have seen diagrams of the Hopf link and King Solomon's knot in which the components cross each other, but no component crosses itself. Find a diagram of the Whitehead link which has this property.

A 2–component link is called **splittable** if it has a diagram where one component doesn't cross over or under the other component. Informally, a splittable link is one that has two pieces which aren't 'linked together.' One of the links in Task 3.6.1 is splittable. Here is another example.

Task 3.6.3: Show that
this link is splittable.

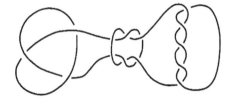

Some interesting 2–component links come from the *double* of a knot diagram.

Start with any knot diagram.

Draw a second strand parallel
to the original. The resulting
link is called the **double** of the
original link diagram.

Task 3.6.4: Show that the link created above is King Solomon's knot. Then find a 2–crossing diagram whose double is splittable.

Task 3.6.5: Find the double of a few diagrams of the unknot. Try to find some rules for when the link is splittable.

The following operation is called a **crossing elimination**. The idea is to cut the strands at a crossing, and reconnect them in a different way so as to eliminate the crossing. There are two different ways to do the reconnection.

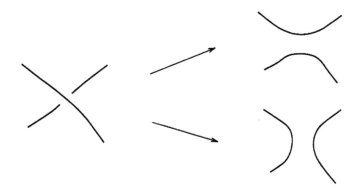

As an example, we apply the above operation to the circled crossing in this diagram of the trefoil:

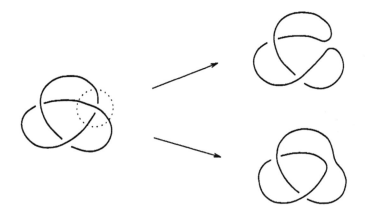

Task 3.6.6: In the example above, we started with a knot diagram. One of the new diagrams represents a knot and the other represents a link. Does this always happen? If 'yes,' explain why. If 'no,' find the other possibilities.

Task 3.6.7: In the example above, one of the new diagrams represents the unknot. Does this usually happen?

3.7 Linking number

Now we develop a way to measure how 'linked together' the components of a link are. This will be based on the *linking number* of a link. The linking number will also be useful for distinguishing between different links.

Start with any link diagram.

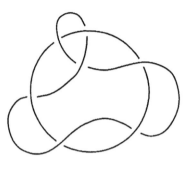

Orient each component.

At each crossing of *different* components, assign either +1 or −1 to that crossing according to these diagrams.

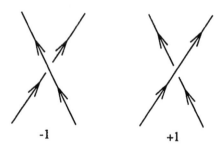

−1   +1

Here are the +1's and −1's for the example above. Note that −1 goes with a left–hand crossing, and +1 goes with a right–hand crossing.

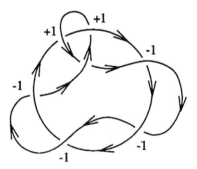

The final step is to add up all the +1's and −1's, divide the total by 2, and take the absolute value of the result. That number is called the **linking number** of the link. In the above example, the sum of all the crossings is $-1 - 1 - 1 - 1 + 1 + 1 = -2$, dividing by 2 gives -1, and taking absolute value we get 1. So the linking number is 1.

Task 3.7.1: Find the linking number of the links below. Do the same for the 2–component links shown at the beginning of the section.

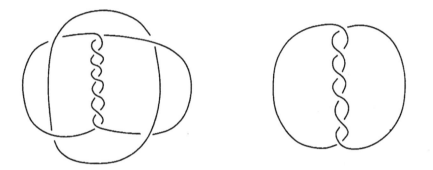

Task 3.7.2: One of the steps in finding the linking number is to add up a bunch of +1's and −1's. Is that sum always an even number?

An Obvious Question: What is linking number good for?

A Less Obvious, but More Important, Question: Does linking number make sense?

We will answer the second question first.

The point of the second question is: we use the link diagram to compute the linking number. If we used a different diagram of the same link, would we get the same answer? We hope that the answer is 'yes,' but since there are so many complicated ways to rearrange a link, it is not obvious that all those different diagrams will give the same linking number. Fortunately, we can make use of an important theoretical tool called the *Reidemeister moves*.

In 1927 Kurt Reidemeister proved that any rearrangement of a knot or link is actually built from three simple steps called the **Reidemeister moves**. The three Reidemeister moves are:
Put in or take out a left–hand or right–hand kink:

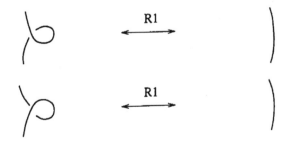

Slide a loop over or under a fixed strand:

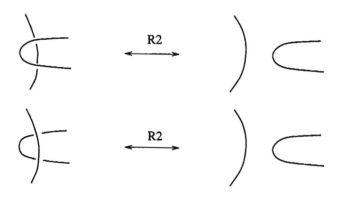

Slide a strand past a crossing:

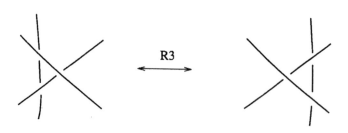

Each of these moves affects one small part of the knot, and leaves the rest unchanged.

The two pictures we drew for R2 can actually be thought of as the same. You can change one picture into the other by turning it upside down, straightening one strand, and bending the other. For R3 we only need to draw one set of pictures because we can think of it either as the 'bottom' strand moving to the right, or as the 'middle' strand moving up and to the left, or as the 'top' strand moving down and to the left.

It may seem quite natural to you that any knot or link manipulation can be rewritten in terms of the Reidemeister moves. However, the proof that the Reidemeister Moves are sufficient makes use of some powerful mathematical tools, and it would take us another whole book just to lead up to it. Reidemeister's theorem is one of many statements in mathematics which seem very sensible and 'obvious,' yet which take lots of time and effort to prove rigorously. This extra work to give a rigorous proof sometimes seems odd to nonmathematicians. The reason for it is that mathematics is about absolute certain proof, as opposed to 'proof beyond a reasonable doubt' or 'it seems to work in all the cases we care about.' In this book we are just beginning our study of mathematics, so we are aiming to explore interesting ideas and we do not worry too much about giving proofs of all the things we assert. If you go on to study the subject at a more advanced level then there will be a great emphasis on giving complete proofs of your assertions.

Task 3.7.3: Choose one or two of the manipulations shown in the text of Section 3.2 and rewrite it as a sequence of Reidemeister moves.

Task 3.7.4: Explain knot manipulation and the Reidemeister moves to a friend who doesn't know anything about mathematical knots.

Returning to our previous questions, the Reidemeister moves give us a way to show that linking number depends only on the link, not on the particular diagram of the link. All we must do is show that the linking number is the same before and after each of R1, R2, and R3.

For example, suppose we do move R2, in the case that the two strands come from different components:

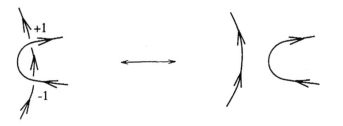

The sum of the numbers is the same (ie, zero) on both sides, so R2 did not change the linking number.

Task 3.7.5: The above example does not completely take care of R2, because the two strands could possibly be part of the same component, and the strands could be oriented differently. Write out the possibilities and check that the move does not change the linking number.

Task 3.7.6: Check that R1 and R3 do not change the linking number.

We have shown that linking number depends on the link, not the choice of diagram of the link. We usually refer to this as, "Linking number is an *invariant* of the link."

Now we can determine a use for linking number. Recall that by switching the crossings in a knot we can get the unknot. Similarly, by switching the crossings in a link, we can get a splittable link. When switching a crossing in a link, there are two possibilities: the crossing could involve strands from the same component, or it could involve strands from different components. The next Task looks at the difference between these operations.

Task 3.7.7: For several two–component links, make crossing changes so that the link becomes splittable. For each one, try to determine:

a) the smallest number of crossing changes needed,

b) the smallest necessary number of crossing changes involving *different* components. For this part you can make as many crossing changes as you want involving a component with itself.

Your answer to part b) of Task 3.7.7 should be the same as the linking number of the link. If you got something different, then go back and check your work! In other words, the linking number of a two–component link measures the smallest number of crossing changes, involving different components, which are needed to make the link splittable.

Task 3.7.8: Check that both of these links have linking number 0. Then make crossing changes *only involving a component with itself* so that the links become splittable.

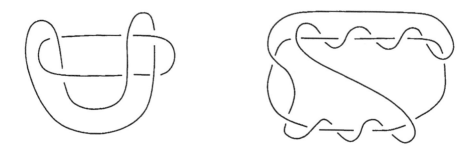

We found that the Whitehead link has linking number 0, the Hopf link has linking number 1, and King Solomon's knot has linking number 2. Since those numbers are all different, we conclude that the links are also different. In other words, linking number can tell the difference between the Whitehead link, the Hopf link, and King Solomon's knot. This is significant, because up to this point we did not actually have any effective way of showing that two knots or links were different.

Task 3.7.9: Draw two different links with linking number 3. How can you be sure that your links are different?

Since a splittable 2–component link has linking number 0, one cannot use linking number to determine that the Whitehead link isn't splittable. So, linking number is an imperfect way of distinguishing between links. In the next section we will discuss another way of distinguishing between various knots and links. This method is also imperfect, but it succeeds in some cases where linking number fails.

3.8 Coloring knots and links

A knot or link diagram is called **3–colorable** if it is possible to do the following:

- color each strand of the diagram,

- use a total of 3 colors,

- at each crossing, all the strands are the same color, or all the strands are a different color. In other words, you can't have two strands the same color and the third a different color.

Here are two 3–colored diagrams. We use the numbers 1, 2, 3, to color the strands. While we use numbers to 'color' the strands (this is a black–and–white book!), you will probably get better results if you use colored pens.

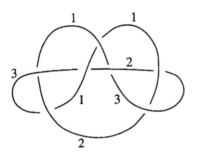 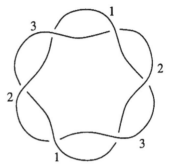

Task 3.8.1: 3–color these diagrams.

Task 3.8.2: Show that these diagrams cannot be 3–colored.

Task 3.8.3: Draw a diagram with between 10 and 15 crossings and try to determine if it can be 3–colored. Given a diagram with many crossings, describe how you would determine if it could be 3–colored.

We see that some diagrams can be 3–colored, and some cannot. However, we are more concerned with studying knots, not just knot diagrams. Fortunately, we have the following fact:

3–Coloring Fact. Given a knot, either every diagram of that knot can be

3–colored, or no diagram of that knot can be 3–colored.

We will establish this fact by using the Reidemeister moves. We must show that if a diagram is 3–colorable, then after doing a Reidemeister move it is still 3–colorable. Here is R2.

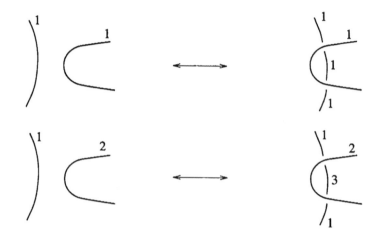

Note that there were two cases to consider, depending on how the strands were initially colored. Note also that the parts of the strands which 'go to the rest of the knot' cannot be changed, because those strands are involved in other crossings of the diagram.

Task 3.8.4: Check that 3–colorability is not changed by the Reidemeister moves.

Task 3.8.5: Explain why the unknot isn't 3–colorable.

Since the trefoil is 3–colorable, and the unknot isn't 3–colorable, we conclude that the trefoil isn't the unknot. It is important to understand the significance of that statement. We have repeatedly said that various knots are different, but this is the first time that we actually *proved* that two knots are not the same. Up until now, a skeptic could have claimed that all the knot diagrams we have drawn might actually represent the unknot, it's just that we haven't been clever enough to rearrange them to get rid of all the crossings. We now know that any 3–colorable diagram doesn't represent the unknot. Of course, this method isn't perfect. For example, the figure–8 isn't the unknot, but it can't be 3–colored.

We must make a slight modification concerning 3–colorability of links. A 2–component splittable link, drawn with the components not crossing each other, can be colored by making one of the components one color, and the other component another color. *This counts as a 3–coloring, even though it only uses 2 colors.* In particular, any splittable link is 3–colorable.

In Task 3.8.2 you showed that the Whitehead link isn't 3–colorable. Since a splittable link is 3–colorable, we conclude that the Whitehead link isn't splittable. It is interesting that linking number failed to show that the Whitehead link is splittable, but 3–colorability succeeded.

3.9 Notes

Note 3.9.a: The diagrams below represent the same knot. The diagram on the left is drawn with as few crossings as possible, and you must switch 3 crossings in order to get a diagram of the unknot. The diagram on the right has more crossings, yet it can be turned into a diagram of the unknot by switching only 2 crossings. This example indicates some of the difficulty of finding the unknotting number of a knot.

3.10 Catalog of knots

Knots are named in terms of crossing number and position in the catalog. For example, the 6_2 knot, pronounced "the six-two knot" is the second knot with 6 crossings. There is no logic to the order in which we list the knots with a given crossing number. The first published catalog of knots listed them in a certain order, and all subsequent catalogs have used the same arbitrary order.

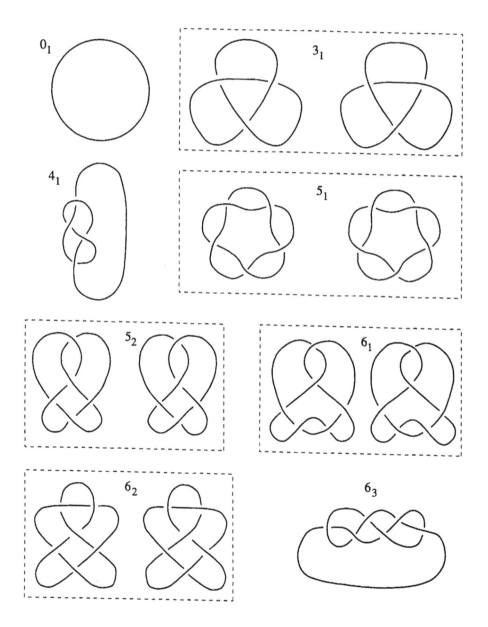

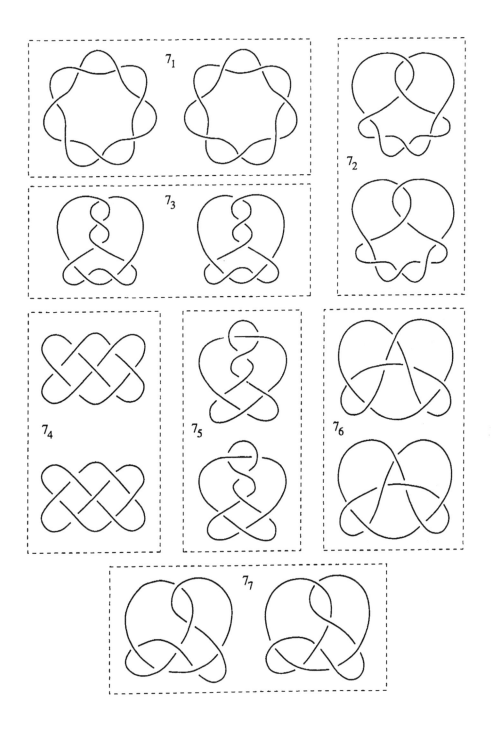

4

Projects

4.1 Ideas for projects. These projects involve applying methods from this book to topics which are not usually considered to be part of mathematics, or they involve collecting together our old work in an interesting way.

These topics appeal to a general audience, so you should include both a general introduction to the subject, and also an introduction to the mathematics you will use in your analysis.

Surfaces as sculpture. Many works of modern sculpture can be identified as surfaces we encountered in Chapter 2. You can find $v - e + f$ for the surface, count boundary curves, determine if it is one–sided or two–sided, and identify it on our list of surfaces. Use your creativity to find other interesting aspects to analyze. For example, the sculpture may appear more symmetric than our usual picture of the surface, or there may be some interesting lines drawn on the surface.

Some useful references:

"Mathematical Ideas Shape Sculptor's Work." Barry A. Cipra, *SIAM News*, May, 1990, 24ff.

"Equations in Stone." Ivars Peterson, *Science News* **138**, 1990, 150–154.

"Mathematics in Marble and Bronze: The Sculpture of Helaman Rolfe Pratt Ferguson." J.W. Cannon, *The Mathematical Intelligencer* **13**, 1991, 30–39.

"Two Theorems, Two Sculptures, Two Posters." Helaman Ferguson, *American Mathematical Monthly* **97**, No. 7, 1990, 589–610.

"3D Mathematics in Wood and Stone." Michael Haggerty, *Displays on Display, IEEE Computer Graphics and Applications* **11**, No. 5, 1991, 7–10.

"Sculptures of John Robinson at the University of Wales, Bangor." Ronald Brown, *The Mathematical Intelligencer* **16**, No. 3, Summer 1994.

The Visual Mind, M. Emmer, Ed., MIT Press, Cambridge, MA, 1993. (This book has several articles showing interesting sculpture.)

Practical knots. For centuries knots have been used for practical and aesthetic purposes. Find a class of knots which interest you and do the following:

– Explain the history and the uses of the knots.

– Give diagrams showing how to make some of the knots.

- Give background information on mathematical knots so that an untrained person interested in practical knots will be able to appreciate the mathematics.

- Analyze the knots mathematically.

You will need some creativity when analyzing the knots. First you must connect any dangling ends in order to get a mathematical knot. If there are more than two loose ends, then this can be done in several ways; some combinations will give knots, and others will give links. Each can be analyzed using the methods of Chapter 3. For example, is it alternating? prime? composite? a link, and if so, how many components? is it 3-colorable?

Some interesting knots:

Sailor knots: Sailors make use of many interesting and practical knots. Look in the 'sailing' section of your local library.

Boy Scout knots: There are dozens of knots in the Boy Scout handbook.

Ornamental knots: Beautiful and intricate knots appear in many Asian cultures. Learn to tie some fancy knots and include them with your project.

Celtic knotwork. These knots are not made from string, but are carved in stone, drawn on paper, and cast in metal. These can be analyzed as described in the *Practical knots* project. Look up 'Art, Celtic' in the library. *The Book of Kells* is a particularly rich source of intricate knotwork.

The family tree of knots. If you switch one crossing of a knot diagram then you get another knot. If you switch a different crossing then maybe you get the same knot, maybe not. Make a big chart showing the 15 or so simplest knots, with arrows pointing to the knots you get when you switch various crossings. You do not have to limit yourself to the knots shown at the end of Chapter 3. First you will have to do the work of finding what you get when you switch each crossing of each knot. Then you will have to find a good way to represent all that information. Also write a few pages giving general information about knots and explaining the purpose of your chart. Note: you will have to decide if it is better to use just one knot out of a mirror image pair. Doing this will allow you to show more knots, but it may not capture the whole story. You can explain your choice in the paper accompanying your chart.

Graphs, knots, and surfaces in art. By using a little creativity you can find many examples of graphs, knots, and surfaces in many paintings and other works of art. See the above descriptions for some ideas for analyzing the objects you find.

Symmetry and aesthetics of graphs, knots, and surfaces. What does it mean to give a 'nice' picture of a graph, knot, or surface? This was very briefly addressed following Task 3.3.10.

Illustrate a mathematical topic. Use your artistic ability to create an interesting picture/sculpture/something else related to the mathematics you have been studying.

Graphs, knots, and chemistry. Graphs: Hydrocarbon compounds are often drawn as if they were graphs. In fact, the problem of listing all hydrocarbons with a given number of carbon atoms can be rephrased in terms of listing graphs with certain properties. Knots: Chemists have synthesized knotted organic compounds, and knotted DNA occurs naturally. See *The Knot Book* [KB] for more details and some references, and also,

"Lifting the Curtain: using topology to probe the hidden action of enzymes," by De Witt Sumners, *Notices of the AMS*, **42** (5) May 1995, p 529.

4.2 Mathematical topics

These projects extend and build upon the material in this book.

Games on Surfaces. Many popular games are played on part of the plane. Some of those games can be modified to make interesting games on other surfaces.

- If you glue the opposite edges of a checkerboard, the result is a torus. Is it possible to modify the rules of checkers to make an interesting game on the torus?

- Is there an interesting 3–player game similar to checkers on a hexagon with opposite edges glued?

- Glue opposite edges of a tic–tac–toe board. Does the game now become interesting? How about 4–in–a–row?

   The Shape of Space [SoS] discusses torus chess and torus tic–tac–toe.

   An interesting game to analyze is **sprouts**, a game invented by Conway and Paterson. Here are the rules.

a) Start with some number of 'spots.'

b) A legal move is to draw a curve connecting any one spot to any other spot (including itself), and then putting a new spot in the middle of that curve.

c) No spot can have more than 3 lines coming out from it.

d) The winner is the last person to make a move.

   Here is a sample game of 2–spot sprouts:

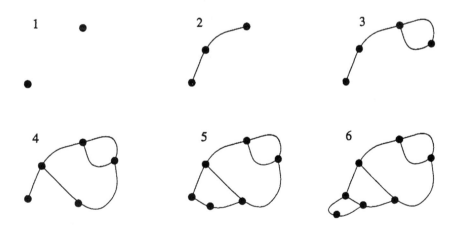

For a given starting number of spots, either the player who moves first, or the player who moves second, can guarantee a win by perfect play. Which player is able to force a win depends on the initial number of spots.

- Analyze 1–, 2–, and 3–spot sprouts. Your analysis will be greatly simplified if you recognize when a position is equivalent to a position you have seen before. Also, it is helpful to think of the game as being played on the sphere, as opposed to the plane. For example, in 2–spot sprouts, there are only two inequivalent first moves.

- Suppose you wanted to play sprouts on the torus or the double torus. How should the rules be modified? Does the same person win?

The game of sprouts is discussed by Martin Gardner in his July 1967 column in *Scientific American*. That article is reprinted in his book *Mathematical Carnival*.

The double torus. In Chapters 2 and 3 we made an extensive study of which graphs and knots can be drawn on the torus. In this project we find the corresponding results for the double torus. Some good things to do:

- Find useful representations of the double torus as a polygon with sides glued. For some ideas see Task 2.10.4 and the standard form given in the Notes to Chapter 2. Note that 'standard form' does not mean 'most useful form.'

- Which graphs K_n can be drawn on the double torus? Draw those that are possible, and show that the others are impossible.

- Repeat the above for $K_{n,m}$ (this is more difficult and can be omitted).

- Find the regular graphs on the double torus.

- Find some knots and links which can be drawn on the double torus. The figure–8 knot is one example.

- Find some knots which are the boundary of a double torus with one hole.

One–sided surfaces. In this project you continue the study of one–sided surfaces begun in Chapter 2. The one–sided surfaces are also known as the 'nonorientable surfaces.' This is discussed in the Notes at the end of Chapter 2. It would take considerable time and effort to make a complete study of one–sided surfaces, so you should concentrate on the parts you find most interesting.

Recall that we say a surface is **closed** if it has no boundary and it can be written as a finite cell complex. The projective plane is defined to be the closed surface obtained by gluing the boundary of a disk onto the boundary of a Möbius strip. The Klein bottle is defined as the connected sum of two projective planes.

- Show that the square–with–sides–glued pictures of the projective plane and the Klein bottle given at the end of Section 2.6 are correct. That is, show that the given form of the projective plane can be cut into a disk and a Möbius strip, and the flat Klein Bottle can be cut into two Möbius strips.

Taking the connected sum of a surface with a projective plane is sometimes called **adding a crosscap**. The projective plane could then be called 'a sphere with crosscap.'

A sphere with crosscap:

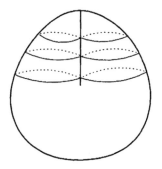

  – Discuss the accuracy of the crosscap representation of the projective plane.

   In order to represent a one–sided surface as a 3–dimensional object, it is necessary to have the surface pass through itself. The crosscap picture is an example. However, if the surface has a hole in it, then it can be drawn without intersecting itself. The simplest example of this is the Möbius strip: it is the projective plane with one hole.

A Klein bottle with one hole:

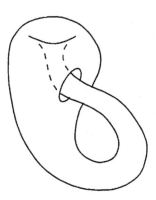

   If you glue a disk onto the hole, then you get the commonly given picture of the Klein bottle. Its shape suggests the name 'bottle,' but you can't put much in a bottle that doesn't have an inside!

   We mentioned in Chapter 2 that the projective plane is the building block for the one–sided surfaces: any one–sided surface is the connected sum of some number of projective planes, with some number of holes. We use $v - e + f$ and the number of boundary curves to identify one–sided surfaces.

  – Show that for the projective plane $v - e + f = 1$, and for the Klein bottle $v - e + f = 0$.

  – Explain why taking the connected sum of a surface with a projective plane decreases $v - e + f$ by 1.

  – Explain why cutting a hole in a surface decreases $v - e + f$ by 1.

We write $A\#B$ for the connected sum of the surfaces A and B, and we denote the projective plane by P and the Klein bottle by K. So $K = P\#P$, and the closed one–sided surfaces are P, $P\#P$, $P\#P\#P$, and so on.

- Make a chart of the one–sided surfaces, labeled with the value of $v - e + f$. You can't draw them, so your chart will contain entries like "$P\#P\#P$ with one hole, $v - e + f = -2$."

- Explain how to tell if a cell–complex represents a one–sided surface.

- Identify these surfaces:

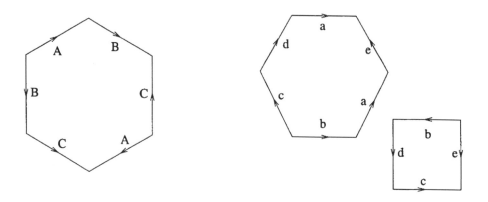

- Explain why the connected sum of a one–sided surface with a two–sided surface is one–sided.

- Let T denote the torus. Which one–sided surface is $T\#P$?

- Determine which graphs K_n and $K_{n,m}$ can be drawn on P and K.

Warning. It is not possible to draw K_7 on the Klein bottle. This is contrary to what you would probably predict. The Klein bottle is the only surface where the method of using $3e \leq 2f$ fails to correctly predict which K_n can be drawn on the surface. Nobody knows a simple reason why the Klein bottle should be the one exception to the rule.

- Find some regular graphs on the projective plane and the Klein bottle.

Space Tyrant mazes. An evil galactic tyrant from outer space has trapped you in a giant maze. You have barely enough room to stand up in the corridors as you slowly explore your cold prison. There are dancing blue lights at the intersection of corridors and along some of the passageways; in their dim light you can see only a few feet ahead. You wander the maze, but find no entrance or exit. You struggle to scratch the wall, but can make no mark, and you have nothing to leave on the floor to make a trail. You often come to places that look familiar, but since you cannot leave a mark you don't really know if you have been there before.

The evil tyrant offers you the following deal: make a graph with the dancing blue lights as vertices and the connecting corridors as edges. If you can identify which graph this produces, then you will be set free.

After a bit of work you realize that the space tyrant's task is impossible. You can't even determine how many vertices the graph has! Convince yourself that it is impossible to distinguish between the following space tyrant mazes:

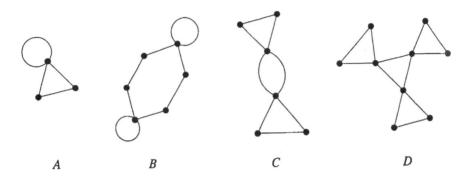

A   B   C   D

The galactic tyrant refuses to obey the usual rules, so his graphs may contain loops and multiple edges.

You confront the space tyrant with the impossibility of his task, and he is impressed that a pitiful human could be so clever. He agrees to modify his offer by granting you one request. In this project you attempt to learn enough about these mazes so that you can make an intelligent request.

For the rest of this project, all terms such as 'equivalent' and 'indistinguishable' are meant in terms of space tyrant mazes.

- Pick a few graphs and for each find a few space tyrant mazes which are indistinguishable from that graph. Devise a method of producing lots of mazes which are equivalent.

- What mazes don't have any other mazes which are equivalent to it?

Here are some ideas to help you find a way out of your terrible predicament:

- What if you were allowed a bunch of different markers which you could leave around the maze? This way you would occasionally know when you returned to a spot. You could then pick up the marker and put it somewhere else. How many markers would you need?

- What if you were allowed to write on the walls? You would know when you returned to a place you had been before, although you would not be able to pick up the mark and leave it elsewhere. How many marks would you need to make?

Covering graphs. This project uses some ideas from the project on Space Tyrant mazes.

A graph is a **covering graph** of another graph if it can be folded up and placed on top of the other graph so that edges lie on top of edges and vertices lie on top of vertices of the same order. We will use the graphs in the Space Tyrant project as examples.

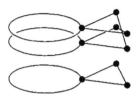

Graph B shown above is a covering graph of graph A:

You should draw pictures showing that graphs C and D shown above are also covering graphs of graph A. For the sake of completeness, a graph is considered to be a covering graph of itself, so graphs A, B, C, and D, are all covering graphs of graph A.

Graph B is called a 2–fold cover of A because every part of graph A has two parts of B lying atop it. Equivalently, when you fold up graph B it lies in two layers on top of A. You should draw pictures to show that C is also a 2–fold cover of A, while D is a 3–fold cover of A. You should convince yourself that all covering graphs are n–fold covers, meaning that when you fold up the covering graph it forms the same number of layers everywhere, as opposed to forming a different number of layers in different places.

- Explain why two covers of the same graph are indistinguishable as space tyrant mazes. (The graphs A, B, C, and D illustrate this.)

- If two graphs can't be told apart, does it follow that they are both covering graphs of the same graph? For example, C and D can't be told apart, and both are covering graphs of graph A.

- Find some graphs which don't cover any other graphs. Are these the same as the mazes which don't appear the same as any other maze?

Does this give a method of outwitting the space tyrant? If you thought of using covering graphs in the Space Tyrant project, would the following request assure that you could determine the true shape of the maze?

'Mr. Tyrant, I request that my maze not be a cover of any other maze."

The evil tyrant is unlikely to grant that request, for the perfectly good reason that he has already constructed one maze for you, and if it happens to cover some other maze, there is no reason he should go to the expense of building you another. Actually, you would be lucky if he denied that request, because it does not guarantee you success. I'll leave it to you to figure out why.

Knots and graphs. This project investigates a way to use graphs to study knots. Here is how to associate a graph to a knot. The graph produced is called a **signed graph** because each edge has a + or − associated to it.

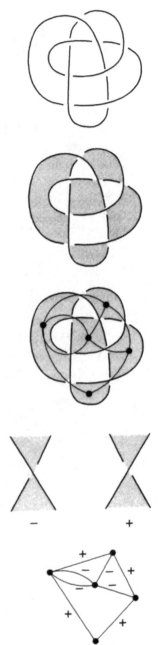

Start with any knot diagram.

Two–color the map determined by the projection of the knot.

Put a vertex in each colored region, and connect vertices by an edge through each crossing of the diagram.

Write + or − next to each edge as shown in these pictures.

This is the signed graph associated to the knot shown above.

Some ideas:

- Show that the above process can be reversed. That is, show how to take a signed graph and produce a knot, and explain why the process, knot → graph → knot, gives back the same knot you started with.

- Determine how the Reidemeister moves work in terms of the graph. For example, here is R1 in the case of a left–hand kink.

In other words, R1 either adds or deletes a loop or a vertex of order 1. Note that there are two cases because there are two possibilities for which side of the strand is colored.

- How does the signed graph of a diagram compare to the signed graph of its mirror image?

- How does the knot associated to a signed graph compare to the knot associated to the dual of the graph? Note: you will first have to decide how to deal with the + and − when you take the dual.

- Analyze the knots associated to cyclic graphs.

- Analyze the knots associated to trees.

- Analyze alternating knot diagrams.

- Analyze the knots associated to graphs which are their own dual.

- Investigate the connection between knot projections and (unsigned) graphs. Is this of any use in finding a catalog of knot projections?

Knot polynomials. See Chapter 6 of *The Knot Book* [KB].

Bibliography

[AC] *Applied Combinatorics*, by Alan Tucker, John Wiley & Sons, 1995.
A good introduction to combinatorics. Has a reasonable exposition of the Polya enumeration formula. A variation of that formula can be used to count the graphs with a given number of vertices.

[E] *Ethnomathematics: A Multicultural View of Mathematical Ideas*, by Marcia Ascher, Brooks/Cole, 1991.
Presents an interesting account of the mathematical sophistication of 'primitive' people.

[F] *Flatland: a romance in many dimensions*, by Edwin Abbott, Dover Books.
An amusing story of a 3-dimensional being who visits a 2-dimensional world. More social commentary than mathematics. An excellent update, without the social commentary, appears in *The Shape of Space*.

[GS] *Groups and Symmetry: a guide to discovering mathematics*, by David W. Farmer, American Mathematical Society, 1996.
A book in the same style as *Knots and Surfaces*.

[GT] *Graph Theory*, by Frank Harary, Addison-Wesley, 1969.
A classic introductory book on graph theory. It begins with lots of definitions, so don't hesitate to skip to the middle and then look back for the terms you need.

[KB] *The Knot Book*, by Colin Adams, W.H. Freeman, 1994.
An introduction to knots. The book elaborates on many of the topics briefly mentioned in *Knots and Surfaces*. A good source of ideas for further study. Mentions many unsolved problems.

[MS] *Mathematical Snapshots*, by Hugo Steinhaus, various publishers.
Short chapters on a variety of mathematics, written for a general audience. All of it is interesting, and two or three of the chapters are relevant to this book.

[SoS] *The Shape of Space*, by Jeffrey Weeks, Marcel Dekker, 1985.
An excellent book. If you liked *Knots and Surfaces*, then you should be able to understand and appreciate most of *The Shape of Space*.

[SSS] *Shapes, Space, and Symmetry*, by Alan Holden, Dover Books, 1991.
Pictures of hundreds of 3-dimensional symmetric shapes. Lots of regular and semi-regular solids.

[VM] *The Visual Mind*, Michele Emmer, Ed., MIT Press, 1993.
A collection of 36 papers dealing with mathematical aspects of art. A few are relevant to this book. Several pictures of interesting sculpture.

Index